Contributions to Economics

For further volumes:
http://www.springer.com/series/1262

Michael Vogelsang

Digitalization in Open Economies

Theory and Policy Implications

Physica-Verlag

Michael Vogelsang
University of Wuppertal
European Institute for International Economic Relations (EIIW)
Rainer-Gruenter-Straße 21
Campus Freudenberg
42119 Wuppertal
Germany
vogelsang@euroeiiw.de
vogelsang@wiwi.uni-wuppertal.de

Diese Arbeit wurde von der Bergischen Universität Wuppertal, Schumpeter School of Business and Economics, im Jahr 2009 als Dissertation angenommen

ISSN 1431-1933
ISBN 978-3-7908-2391-2 e-ISBN 978-3-7908-2392-9
DOI: 10.1007/978-3-7908-2392-9
Springer Heidelberg Dordrecht London New York

Library of Congress Control Number: 2009941526

Cover design: SPi Publisher Services

Printed on acid-free paper

Physica-Verlag is a brand of Springer-Verlag Berlin Heidelberg
Springer-Verlag is part of Springer Science+Business Media (www.springer.com)

Thanks to lovely Eva, vivid Patrick, and keen-minded Christopher Philipp.

Preface

Economic cycles are deeply anchored in human consciousness: seasons determine the time to seed and to harvest, cycles of 14 years are described in the Bible.

On the contrary is the linear perception of life: birth and death are the starting and end-points – at least of our body. But, when alternating news about births and deaths are received in a short interval of time, this induces the impression of a cycle again. During the time I worked on this dissertation I very gratefully cheered the births of my sons Patrick and Christopher Philipp Vogelsang, as well as of their cousin Frida. I will remember always my brother Manfred and my friends Andreas Strömel and Dr. Torsten Gager, who died too early, as well as my grandmother Friederike, who managed to reach nearly 98 years.

When my grandmother was born digitalization as known today was unimaginable. Although the telegraph and telephone allowed worldwide communication, it was inconceivable that a photo, music or a document could be transfered overseas within a second.

But today these are reality. Technological evolutions such as digitalization have also changed our economic perception of time from the circular spring, summer, autumn, winter rhythm to a linear cognition. In economics, some more antagonism exists: between static and dynamic theory and, between macroeconomic and microeconomic views. The idea of this thesis to analyse the links between a sector-specific view and macroeconomics is surely a result of the attractiveness of antagonism.

I owe gratitude to some strong, clever and prudent persons for providing support and motivation to carry on with this work. First, my wonderful wife Eva, who also has the potential to become a PhD student in economics, and my mother for always being in a good temper. I express my gratitude to my PhD adviser Paul J. J. Welfens, a packet of innovative thinking and encouraging energy to bring forward the science of economics. My friend and colleague Zori Kutlina: We explored together the depths of econometrics. Then, the knight of liberal thinking, Martin Keim. Special thanks to Dirk Wentzel and Dennis Davis, who made possible a stay at Penn State University in the final stage of the work, as well as Kerstin Schneider and Gerhard Arminger, excellent scientists at the University of Wuppertal, who motivate PhD students by showing open-minded interest in their research results. Last but importantly I thank my colleagues at the European Institute for International Economic Relations (EIIW) and the Schumpeter School of Business and Economics at the University of Wuppertal for their support and motivation, and Dhiren Bahl for the excellent copyediting.

Wuppertal
Spring 2009

Michael Vogelsang

Contents

Part II Economic Theory

List of Figures

List of Tables

Part I
Characteristics of the Digital Economy

Chapter 1
Introduction

Long economic waves, also called Kondratiev waves, are related to inventions and technologies, which change not only production processes but also the way of living. After the steam, steel, electricity, and petrochemical revolutions, network-based digitalization is the driving force today on the stage of business and private life. The Internet as a mass phenomenon, which began in the 1990s, allows easy and cheap connection of information technology systems. The next steps of technological progress as well as innovations in business and social life rely on information technology. As Nefiodow (2001) puts it: The 6. Kondratiev wave, which could be health and bioengineering, will be based on today's 5. Kondratiev wave, which is information technology.

A technological prerequisite of these developments is digitalization, which means the expression of information in strings of 0 and 1, called binary or digital strings. They are omnipresent in today's life: music, videos, also the control systems in cars, on-board computers in airplanes, encoded human DNA, and, importantly, emails and web pages are encoded in 0 and 1.

Digital strings are not visible, but affect all economic segments. That is in common with electricity, another technological revolution. Furthermore, both can be stored: electricity in a battery; a digital string, e.g. an email, on a medium such as a CD. But only the network between producer and user allows the instant exchange and use of digital goods and electricity respectively. This shows that the phenomenon of digitalization has to be attributed to the interplay of several products – just as power plants and energy networks have to be considered together to explain the international trade of electricity.

In this research, digitalization has been separated into three segments which have their own different economic characteristics: networks, IT services, and digital goods. This separation follows the technological structure of the Internet, which is explained in more detail later.

As Fig. 1.1 shows, networks are crucial for the delivery of IT services and digital goods. IT services control and administer networks and digital goods. Individual software is also part of IT services. Digital goods are those such as music and video files which can be exchanged via the Internet. At the intersection of the three areas, 'Internet applications' represent complex products, which build on networks, services and digital goods. This explains why the 'Internet' is often used as a synonym for digitalization, although this is technically imprecise.

M. Vogelsang, *Digitalization in Open Economies*, Contributions to Economics,
DOI 10.1007/978-3-7908-2392-9_1,

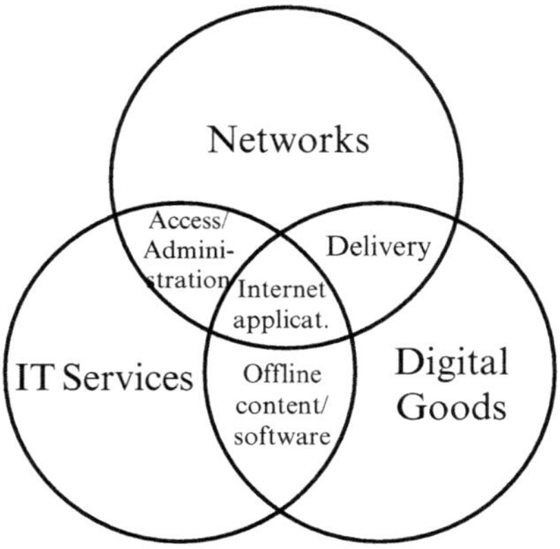

Fig. 1.1 Relationships between digital goods, IT services and networks

Network-based digitalization is also a global phenomenon as the Internet network is available worldwide and the costs of transporting a binary string over the Internet do not rise significantly with distance. Although the 'global village' is an oft-used buzzword, this does not imply that there are no different, country-specific patterns in digitalization. Therefore, this study refers to macroeconomic models of open economies and extends them for networks, IT services and digital goods. The two main questions of this research are:

- Are there any interdependencies between macroeconomic variables and networks, IT services, and digital goods?
- What are the policy challenges of digitalization?

The structure of the study is as follows: An overview of the development of digitalization, the characteristics of networks, IT services and digital goods, recent developments in these segments and the results from the research analysing productivity effects of information and communication technology are presented in Chap. 2.

In Chap. 3, theoretical foundations are presented. These are microeconomic aspects of networks, IT services and digital goods, especially two new approaches regarding the dynamics of two-sided markets and IT administration rights. To introduce the concepts of open economies, theories on internationalization are also presented.

In Chaps. 4, 5 and 6 for each segment, networks, IT services, digital goods, a theoretical general equilibrium model is developed to analyse the interdependencies between macroeconomic variables, digitalization, and specific business strategies. The models stress the international aspects and take into account the new approaches of two-sided markets and IT administration rights, which have been introduced before. The following questions are focused upon:

- Chapter 4: Do **networks** have an influence on international trade (and vice versa)?
- Chapter 5: Do macroeconomic factors (e.g. factor endowments) influence the structure of **IT Service** trade?
- Chapter 6: Is the size of a country important for the internationalization of **digital goods**? Is there a pattern, e.g. is trade, or are acquisitions preferred?

Finally, in Chap. 7, the models for open economies with networks, IT services, and digital goods respectively are compared and first conclusions for economic policy chapters are drawn.

The following chapters focus on economic policy in the digital age and real world politics are compared to the predictions of the theoretical models. In Chap. 8, political perspectives are discussed, which reflect the impact of digitalization on general objectives. Employment policy, competition policy and regulation, data privacy and safety are some of the topics analysed. Important institutional players such as the WTO, EU and other institutions, such as ICANN, which are responsible for the governance of the Internet are introduced. Chapter 8 prepares the ground for Chap. 9, which is more specific on economic policy for networks, IT services, and digital goods.

In Chap. 9, the focus changes to segment-specific policy approaches. It is organized according to the triad of networks, IT services, and digital goods, which have been introduced before.

Regarding networks, there are regulatory instruments and access and interconnection policies. Two sections are also dedicated to the most recent debate about network-neutrality and the political impacts of new generation networks (NGN). Regarding IT services, the importance of standards is emphasized. Then sections about national perspectives, IT services and e-governement, and the global trading approach (GATS and IT-service) follow. Regarding digital goods, the topics of intellectual property rights (IPR), trade policy, and the political consequences of the two-sided market approach are accentuated.

Having separated digitalization into networks, IT services, and digital goods so that macroeconomic general equilibrium models and sector-specific policy analysis become applicable, a joint analysis is presented starting with Sect. 9.4. In this final part of Chap. 9, convergence issues are analysed. Convergence means that several services can be distributed through different networks, e.g. television is not based any more on broadcast networks only. Convergence changes the traditional assessment of policy instruments.

Finally, Chap. 10 integrates the theory and policy parts. The results from the macroeconomic models are matched with economic policy observations. It is shown that predictions from theoretical models about political strategies are justified and appropriate approaches can be observed on the political agenda. For example, the support from governments for domestic companies is a meaningful political strategy, as predicted from the models with digital goods.

The study concludes with a proposal for a layered-orientated policy concept. This concept refers to basic layer principles of the Internet. It is a suggestion for a new policy approach, which is appropriate to digitalization and avoids immanent contradictions, which are a characteristic of today's service-related policy approaches. As

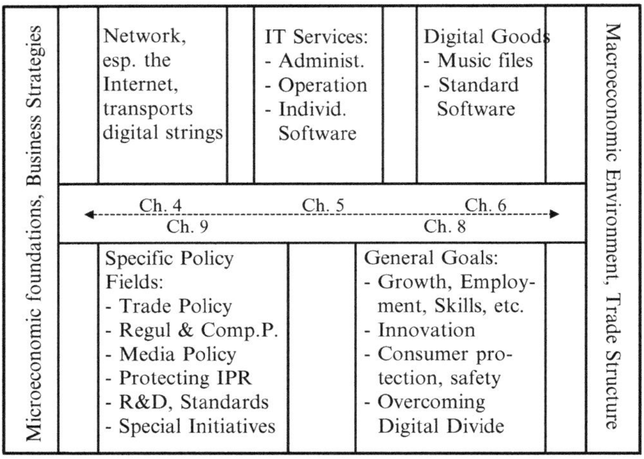

Fig. 1.2 Aspects of digitalization

digitalization and the network expansion bring about so much internationalization also the question has to be solved how to balance the interests of the national and regional policy actors on the one hand and global harmonization approaches on the other.

Figure 1.2 presents the structure of this study graphically. The challenges for a prudent design of economic policy are enormous in the age of digitalization. Digitalization affects all areas of economic and social life. It changes the conditions of fully developed fields such as broadcast regulation. It affects producers, traders, and consumers globally in a very short time. It raises the question of global political coordination, especially regarding the governance of the Internet and convergence issues. And, equally important, today's political set-up prepares the ground for tomorrow's political and economic framework. For example, artificially-intelligent robots which speak, walk, learn to process new information and update themselves via networks are on the starting blocks. Nordhaus (2007) estimates that, at the present rate of improvement in computational ability of about 80% per year, a microcomputer with computing capabilities comparable to human capabilities will cost USD 1 in the year 2035. This upcoming development motivates the need to deal with the process of digitalization, to explore the interdependencies between macroeconomic variables and the international diffusion of networks, IT services, and digital goods today, and to prepare the ground for a consistent international policy model.

Chapter 2
Digitalization

2.1 Overview on Digitalization

2.1.1 History

Digitalization was introduced by Gottfried Wilhelm Leibniz, who developed the transformation from arabic numbers to binary strings. For example, the arabic number 26 could be decoded to the digital string 'I I 0 I 0'.[1] In this research, the adjectives 'digital' and 'binary' are used synonymously for a two-state variable, although 'digital' just means countable (from Latin digit for finger).

Early applications of digital information transfer had been signal techniques (light on/off) for the use of the Morse alphabet, which was invented by Samuel F.B. Morse in 1835. The Morse Code allowed the transport of binary signals by telegraph. The invention of the telegraph 'arose more or less independently from each other in the US, Germany and England, but resting on a common foundation of technical knowledge' (Calvert (2000)). In Europe, Samuel Thomas von Soemmering (1809), Baron Schilling von Canstedt (1820), Carl Friedrich Gauss and Wilhelm Weber (1833) developed telegraphs. Schilling von Canstedt had already used a binary system for signal transmission, whereas the telegraphs of Gauss and Weber used four states. The binary system proved to be less vulnerable to poor signal quality. In the following years, a revised system of the Morse alphabet became standard.

After the telegraph had been invented the technology started to 'go global'. The first fully-functional transatlantic telegraph cable was completed in 1866. A cable from England to India followed in 1872. Pacific Line completed a worldwide net of telegraph lines in 1902. A figure of the telegraph connectivity of 1891 is shown in Appendix 12.1.

A successful telegraph system was characterized by a widespread cable network and digital signal transmission. In the 150 years since the development of the telegraph, Information and Communications Technology (ICT) has evolved rapidly. The

---

[1] $0x2^0 + 1x2^1 + 0x2^2 + 1x2^3 + 1x2^4 = 26$.

M. Vogelsang, *Digitalization in Open Economies*, Contributions to Economics,
DOI 10.1007/978-3-7908-2392-9_2, © Springer-Verlag Berlin Heidelberg 2010

telephone and Electronic Data Processing (EDP) have been invented. The first electronic automatic computer was invented in 1946, and the microprocessor in 1971. The launch of the first personal computers (PCs) is dated variously from 1950 (Simon), to Apple II (1977) or the IBM PC (1981). Nordhaus (2007) estimated that the growth in computer power has averaged 55% per year since the end of World War II. Figure 2.5 shows the costs related to different computing technologies.

The precursor of the Internet came in the late 1960s. A branch of the US Department of Defense developed a network to link university and defence contractors. In the middle of the 1980s, the backbone topology of the Internet was introduced by the NSFNET (National Science Foundation Network) in the US. Regional networks shared access to regional 'supercomputers' which had been connected together via the 'backbone'. The backbone topology directly reflects the cost structure: 'lots of cheap routers are used to manage a limited number of expensive lines' (Mackie-Mason and Varian (1993)). Although the backbone topology has been replaced today by several Internet Exchange Points (IXPs) where Internet Service Providers (ISPs) exchange traffic directly, the word 'backbone' is still used for high-speed connections between providers. The TCP/IP Protocol (Transmission Control Protocol over Internet Protocol) was also introduced by NSFNET in 1986 which is still the foundation of today's Internet traffic.

The 'World Wide Web', which was introduced in 1989, is a system of linked sites with graphical elements, which can be read with a browser software and is often called 'the Internet' today. In 2004, the Morse Code was extended with Δ−−Δ−−Δ by ITU, which represents the symbol @ used in email addresses. So the old Morse and the young Internet technique converge.

2.1.2 *Digitalization and the Internet*

2.1.2.1 Structure of the Internet

Although the differences in speed, technology and comfort between the telegraph and the Internet are obvious, some basic principles are common: transmission of the signal demands a sufficient line quality and the transmitter and receiver must agree on a protocol (a code book) to encode and decode the digital signals. The difference: telegraph operators had to encode and decode words to the Morse alphabet manually. Today, the encoding and decoding are done automatically.

Using the Internet, digital signals are sent in packets and the processing of the received packets takes place digitally in the computer of the receiver. Table 2.1, which is partially based on UNCTAD (2006), shows the layer architecture of today's Internet applications.

In the last row of the figure the economic equivalents are listed. Networks encompass all the physical transmission of data. IT services refer to the administration of the systems. Software is in-between as it can be regarded either as an IT service or as a digital good (see Sect. 2.4.4). Finally, digital goods represent a new class of

Table 2.1 Internet layers

Technical layer	Subject	Economic perspective
Physical network	Binary transmission	
Data link	Ethernet, Wi-fi	Networks
Network (IP)	IP (IPv4, IPv6) etc.	
Transport	TCP, UDP etc.	
Application protocols	HTTP, FTP, etc.	IT services
Application services	Browsers, eMail-clients, streaming media players, etc.	
Content	Text, data, audio, video, etc.	Digital goods

goods with special economic characteristics (see Sect. 2.2.3) which was not available during the era of the telegraph, since information could not be stored digitally at that time. Reproduction, also of cord cards, has been costly.

2.1.2.2 General Purpose Technology

The telegraph and the Internet are both inventions which have led to huge infrastructure investments after their invention. They have also influenced the business strategies of companies which are not part of the telecommunication or information sectors. Today, information technology is considered 'general purpose technology' (GPT, Bresnahan and Trajtenberg (1995)). Following Lee and Guo (2004) the GPT character effects, that ICT

- Enables complementary organizational investments and coordination, which lead to a productivity increase by reducing costs. The complementary investments are important for realizing the productivity benefits of ICT investments (Basu and Fernald (2007)).
- Enables firms to offer new products or services; growing role of convenience, timeliness, quality and variety (Brynjolfsson and Hitt (2000)).
- Induces positive external effects (spillover effects); also enables positive knowledge spillovers, in the sense of Romer (1989).
- Large diffusion in different industries and large variations of product offerings.

2.1.3 Indices of Digitalization

Sciadas (2004) emphasizes that the Information Society was no exception in the lack of a common nomenclature in its early stage. As a result, different concepts are used to measure digitalization. This subsection offers a brief overview of the structure of the most important international benchmarking indices. These are:

The ICT-OI (see Table 2.3) claims to measure the 'infostate', which is defined as the result of 'infodensity' and 'info-use' of a society. The index takes into account

Table 2.2 International benchmarking indices of information society

Abbr.	Name	Provided by
DAI	Digital access index, merged in ICT-OI	International telecommunication union (ITU)
DOI	Digital opportunity index	ITU
ICT-OI	ICT opportunity index, Result of a merger from DAI and Orbicom's index, first published Nov. 2005	ITU
ISI	Information society index	IDC, an advisory company
NRI	Network readiness index	World economic forum,INSEAD
Orbicom	Orbicom digital divide index, merged into ICT-OI	Orbicom / UNESCO

Table 2.3 Structure of the ICT-OI index

Categories	Examples of indicators used
Infodensity	Networks: Main telephone lines per 100 Skills: Adult literacy rates
Info-use	Uptake: Television households Intensity: Broadband Internet subsribers per 100

data of 183 economies and is predominately used to analyse the digital divide (see Sect. 2.3.4).

The DOI (see Table 2.4) focuses more on modern services. In total, ICT-OI uses ten different indicators, the DOI 11 indicators. For a complete list, see ITU and UNCTAD (2007), p.16.[2] Although only the indicator 'mobile cellular subscribers per 100 inhabitants' appears in both indices, they are highly correlated (R-squared equal to 0.94 (ITU and UNCTAD (2007))).

The NRI (see Table 2.5) is claimed to 'access countries propensity and preparation in benefiting from, and participating in, ICT advancements' (Weforum and Insead (2007)):

A detailed list of the indicators used can be found at Dutta and Jain (2004) and Weforum and Insead (2007).

The different (sub-)indices or categories can be used according to the state of development of the country, e.g. an oft-used framework to describe the state of a country regarding its electronic commerce activities is the 'S-curve'. Having time on the x-axis, and the level of e-commerce activity on the y-axis, the S-Curve describes the 'slow' start of preparing the infrastructure, followed by accelerating use, and finally the increasing value added by e-commerce (see OECD (2005)). Fans of linguistic creativity also call the three stages 'e-readiness', 'e-intensity' and 'e-impact'. The S-Curve shows that countries may differ in their state and speed of supplying and using ICT that may favour the use of different information society indices.

[2] An overview of core ICT indicators is provided by UNCTAD (2005).

Table 2.4 Structure of the Digital Opportunity Index

Categories	Indicators used
Opportunity	Percentage of population covered by mobile telephony
	Internet access tariffs as a percentage of per capita income
	Mobile cellular tariffs as a perc. of per capita income
Infrastructure	Households with a fixed-line telephone per 100
	Households with a computer per 100
	Households with Internet access at home per 100
	Mobile cellular subsribers per 100
	Mobile Internet Subscribers per 100
Utilization	Proportion of individuals that have used the Internet
	Ratio of fixed broadband subscribers to total Internet subscr.
	Ratio of mobile broadband subsribers to total mobile subsribers

Table 2.5 Structure of the Network Readiness Index

Categories	'Pillars'	Examples of indicators used
Environment	Market environment	Venture capital availability
	Political and regulatory environment	Laws relating to ICT
	Infrastructure	Telephone lines
Readiness	Individual readiness	Quality of schools
	Business readiness	Extent of staff training
	Government readiness	Government prioritization of ICT
Usage	Individual usage	Broadband Internet subsribers
	Business usage	Extent of business Internet use
	Government usage	Availability of online services

Table 2.6 shows the ranks of the DOI compared with other indices.[3] Although the structure of the indices differs, comparing several information society indicators may lead to common conclusions (Lauri and Hirvonen (2006)). This favours the approach of this study to consider digitalization as a broad concept. IT occurs independently from a value of a single ICT indicator.

2.2 Characteristics of Networks, IT Services and Digital Goods

In this section, the characteristics of digital goods, networks and IT services are described. Related to this are the concepts of ICT and e-commerce which are introduced subsequently.

[3] The data are based on a project of ITU and the Korean Agency for Digital Opportunity and Promotion (KADO). The ranks may differ from original data because numbers in this table express ranks among countries shown here.

Table 2.6 Comparison of DOI with other indices; Data source: ITU (2005)

	DOI	NRI	ISI	DAI	Orbicom
Sweden	1	4	1	1	1
Denmark	2	2	2	2	2
Korea (Rep.)	3	17	13	3	14
Switzerland	4	7	4	10	6
Hong Kong	5	5	15	5	7
Singapore	6	1	9	11	9
Taiwan	7	12	19	6	
Netherlands	8	13	3	4	4
Japan	9	6	14	12	15
United States	10	3	5	8	5
Canada	11	8	7	7	3
United Kingdom	12	10	8	9	11
Germany	13	11	10	14	10
Austria	14	15	6	13	13
Belgium	15	18	12	16	7
Australia	16	9	11	15	12
France	17	16	17	18	16
Italy	18	28	18	17	18
Israel	19	14	16	19	17
Spain	20	20	20	20	19
Czech Republic	21	26	21	21	20
Hungary	22	24	22	22	21
Poland	23	37	25	23	24
Malaysia	24	19	27	25	25
Chile	25	22	23	24	22
Argentina	26	38	26	26	23
Russia	27	34	31	27	30
Mexico	28	33	29	28	27
Turkey	29	31	36	31	28
Thailand	30	23	34	30	33
Egypt	31	32	33	38	37
China	32	27	37	36	35
Colombia	33	35	30	34	32
Venezuela	34	39	32	32	31
Brazil	35	29	28	29	26
South Africa	36	21	24	33	29

2.2.1 Networks

In the early days of the telegraph signals could be sent on one line only, say between two cities. To change to another line (to reach a third city or local destination) at the telegraph office, the operator had to listen to the first signal, write it down, and then manually key the Morse signal on the other line (Standage (2007)).

Then, in the telephone age, networks were built up in a star topology, which means that all users are connected to a central switchboard. This prevents build-up of $N(N-1)/2$ connections which would be necessary to connect N users in a network directly. The voice signal of the Public Switched Telephone Network (PSTN), which is analogue, is used.

Today, the task of interconnection is done by hardware, but the two main components of a network remain the:

- End-user connection; plus
- The exchange points between the lines.

In contrast to the fixed-line age today, mobile and satellite connections replace the copper wires and fibre cables. And in contrast to circuit switching of telephone data networks, packet switching is used instead. This means that the data are split up into packages, labelled with the address of the destination, and sent from node to node. The transmission volume of one cable can be shared, and available bandwidth and delay times of the packets vary with the use of the network. Data networks allow a decentralized topology. Internet Service Providers (ISPs), which give access to the customers, are connected to Internet Backbone Providers (IBPs) which can exchange data directly via a third IBP (Economides (2005)). The exchange points between the different providers are also called ‘peering’, ‘interconnection’ or ‘transit points’ (see Sect. 2.3.1 or Sect. 9.1.2 for an interconnection model).

Today, data networks are also used for transmission of voice (Voice over IP, VoIP; see also Sect. 2.3.4). TV and radio networks also have special characteristics, since they usually only allow a one-to-many connection in contrast to telephone and data networks. For more details about network technology, see Hatfield et al. (2005).

In economic theory, networks (more precisely the service of transmission over a network) are usually considered a homogeneous good, although the ‘quality of service’ of a network can differ regarding speed, reliability, etc. Shy (2002) summarizes which characteristics distinguish network markets from ‘the market for grain, dairy products, apples, and treasury bonds’.

- **Complementarity, compatibility and standards:** Network hardware (e.g. routers, switches, hubs), software and cables have to be compatible, i.e. rely on common standards. Then they can be considered as complementary goods.
- **Consumption externalities:** The utility of a consumer by accessing a network rises when more users use the network.
- **Switching costs and lock-in:** When switching costs are high, users do not change the network easily. They are ‘locked-in’.
- **Significant economies of scale:** Marginal costs get lower and average costs decline when more users access the network.

The basic approach of network economics, which focus on consumption externalities, is presented in Sect. 3.1.1. The (potential) speed or bandwidth of a network is measured by data volume in bits per second. The standard abbreviations used are shown in Table 2.7.

Table 2.7 Speed of networks

1 b/s	bit per second	1 byte	= 8 bits
kbps	kilobit per second	1 KBps	1 Kilobyte per second
mbps	megabit per second	1 MBps	1 Megabyte per second
gbps	gigabit per second	1 GBps	1 Gigabyte per second
tbps	terabit per second	1 TBps	1 Terabyte per second

The factor between the columns of bits and bytes is always 8. The factor between the rows is 1,024, often approximated to 1,000 (=kilo).

2.2.2 *IT Services*

According to the 1993 System of National Accounts (SNA), services are defined as

> not separate entities over which ownership rights can be established. They cannot be traded separately from their production. Services are heterogeneous outputs produced...By the time their production is completed they must have been provided to the consumers (UN (2002), p.7).

IT services are a subset of commercial services, encompassing the provision and administration of IT resources. Regarding the Internet, the necessary IT services are shown in the left row of Table 2.1. In contrast to IT services, ICT services are also additionally related to communication technology. OECD defines that, 'ICT services must be intended to enable the function of information processing and communication by electronic means' (OECD (2005)).

In 2002, the European Communities proposed (WTO (2002) – see also Sect. 9.3.3) that Computer and Related Services (CPC 84),[4] regardless of whether they are delivered via a network, should include all services that provide:

- Consulting, strategy, analysis, planning, specification, design, development, installation, implementation, integration, testing, debugging, updating, support, technical assistance, or management of or for computers or computer systems;
- Computer programmes, defined as the sets of instructions required to make computers work and communicate (in and of themselves);
- Data processing, data storage, data hosting or database services;
- Maintenance and repair services for office machinery and equipment, including computers;
- Training services for staff of clients, related to computer programmes, computers or computer systems, and not classified elsewhere.

Other differentiations are applicable. EITO (2005) mentions consulting, implementation (installing, testing, training, etc.), operations management and support

[4] The EC proposal to WTO means that computer and related services could be listed at a two-digit level of the CPC (Central Product Classification – see Sect. 2.2.4) system, which may prevent long-term negotiations when new computer and related services are brought into the market. On the other hand, audiovisual services are not included in this proposal.

services. Dibbern and Heinzl (2002) point to the functions of IT and introduce the following classification:

- System operations/data centres, e.g. installation, operation and technical maintenance of client/server systems or systems software;
- Telecommunication[5]/networks[6];
- Development and implementation of applications, e.g. development of software, system analysis, project management, data administration;
- Help desk, e.g. advice, support, training, instruction and problem management;
- Planning and management, e.g. integration of business planning and information systems planning, identification of future innovations, conception of system architecture, standards and methods.

Another classification of functions in companies is given by Allweyer et al. (2004). They differentiate between serving processes (organizational aspects), applications (e.g. software) and infrastructure (e.g. hardware, networks, operating systems). A further classification from EBOPS is presented in Sect. 2.4.4. For the proposal of a new classification system of ICT services made by the OECD 'Working Party on Indicators for the Information Society', see OECD (2007a).

An example of an IT service is hosting a website. This means supplying a server (hardware) with a webserver (programme) which sends .html (and other) codes when requested for. Customized software could also be considered an IT service. See Sect. 2.4.4 for more information on the differentiation between digital goods and IT services.

2.2.3 *Digital Goods*

Digital goods are those which are expressed in binary 0 and 1 strings. They can easily be distributed over data networks (e.g. the Internet). Examples of digital goods are music files (e.g. .mp3, .wma, .rm files), mobile ring tones, electronic books (or scientific papers) and standardized software (see Sect. 2.4.4). Many digital goods can also be regarded as information goods, e.g. information provided in a text-file format. According to Quah (2002), digital goods are:

- Nonrival;
- Infinitely expansible; nonrivalry is possible without infinite expansibility, but infinite expansibility always implies nonrivalry; (Howell (2004)).
- Discrete;
- Aspatial[7];

[5] Not part of the definition of IT services used in this research.

[6] See Chap. 2.2.1.

[7] Aspatial: at once everywhere and nowhere.

- Recombinant; this is equivalent to the economic conditions of economies of scope. As disussed by Shapiro and Varian (1998), the value of two combined pieces of information $U(D_1, D_2)$ is higher than the added value of these pieces $U(D_1) + U(D_2)$.

These characteristics imply that the marginal costs of producing digital goods are almost zero. They are costly to produce but cheap to reproduce. They can be reproduced arbitrarily often without loss of quality. When the price exceeds marginal costs, the benefits of price discrimination are apparent (see Sect. 3.1.3). Versions and adaptations can be produced easily, implying new pricing and marketing models. The transportation costs of digital goods are very low compared to those of manufactured goods. They are non-perishable, easy to store, therefore, production and consumption can be separated in location and time. Finally, quick and comprehensive analysis of customer behaviour (Mahnke and Venzin (2002)) is possible.

It should be noted that these characteristics of digital goods describe pure cases. For example, in the case of software, Howell (2004) points out that software is rival in use, as it uses processor time (when it is not available for other processes) and may be restricted in use due to licences. Mixed digital goods also exist, where the digital information is saved on a tradable device. In the literature, some more classifications are used: Kling and Lamb (2000) differentiate between highly and mixed digital goods, e.g. a music CD is a mixed digital good.

2.2.4 Related Topics

2.2.4.1 Information and Communication Technology (ICT)

An analogue apparatus such as a 35 mm film camera is also an ICT good, but it does not belong to a network, or an IT service, and is not a digital good, only these being basically considered here. ICT is a broader concept than digitalization.

Some care is necessary to classify ICT products. First, some products are on a borderline between categories (e.g. software, see Sect. 2.2.4). Second, the structures of the statistical systems, on which the classifications are based, differ. Table 2.8 shows the differences following the OECD definitions.

CPC is the UN's Central Product Classification, ISIC the International Standard Industrial Classification, also provided by the UN. The Harmonized Commodity

Table 2.8 ICT classifications and statistical systems

Category	Based on statistical system
ICT sector	ISIC Rev. 3.1 (see OECD (2005), p. 102)
ICT goods	HS (see OECD (2005), p. 88)
ICT services	CPC (see OECD (2005), p. 17)

Description and Coding System (HS) is provided by the World Customs Organization (WCO) in Brussels. HS is primarily used for trade statistics, but does not classify intangibles.[8]

Therefore, trade statistics of ICT services are estimated within the balance of payments framework (for the definition of services used in this framework, see Sect. 2.2.4). Despite the differences, some conversion keys between the systems can be calculated, but they do not fully convert all products and services without loss of information quality.

Generally, the ICT sector can be categorized into ICT goods and services:

- ICT services (see definition in Sect. 2.2.2):
  - Goods-related (e.g. renting of office machinery);
  - Intangibles (e.g. telecommunications);
- ICT goods (based on HS):
  - Telecommunications equipment;
  - Computers and related equipment;
  - Electronic components (e.g. transistors, inductors, circuits, etc.);
  - Audio and video equipment;
  - Other ICT goods (e.g. cash registers, navigational aid apparatus, electrocardiographs, meters, thermostats).

Ark et al. (2002) differentiate the sectors between ICT product manufacturing (e.g. wire, cables, semiconductors), ICT-producing services (posts and telecommunications, computer and related services), ICT-using manufacturing (e.g. printing and publishing, aircraft), ICT-using services (wholesale and retail trade, financial intermediation, insurance, renting, research and development, professional business services) and less-intensive ICT-using industries (e.g. food products, textiles).

The importance of ICT for economic growth is twofold: as productivity-enhancing 'general-purpose technology' (see Sects. 2.1 and 2.4) and as an input factor. The importance of ICT goods as an input factor is documented by Jorgenson and Vu (2007): they find that the worldwide 3.75 p.a. GDP growth (based on 110 economies) can be attributed to ICT capital (0.42), non-ICT capital (0.86), labour hours (0.70), labour quality (0.37), and total factor productivity growth (1.40) in the period 2002–2004. In the period 1995–2000, ICT capital as an input factor attributed 0.82% points to global GDP growth p.a. A closer look on the contribution of ICT to

[8] The only way to measure imports and exports of software with trade statistics is to track the trade in physical media (e.g. CDs, DVDs). But this approach has limitations (OECD (2005)):

- Bundling of software and hardware overstates the equipment trade and understates the software trade;
- 'Gold Master Problem', when only the original software is transferred internationally and copied inside the country. Then the value of copyright transfer is not accounted for;
- Software which is transmitted via the Internet is not measured by trade statistics.

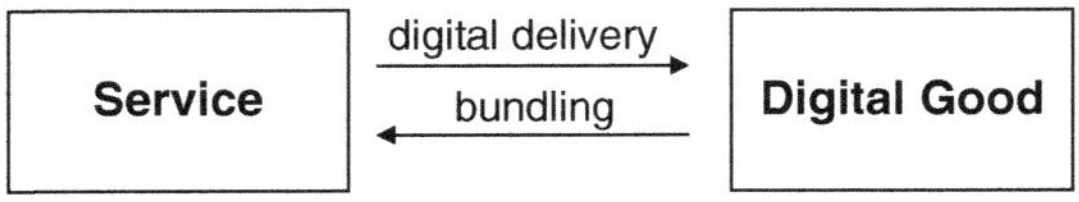

Fig. 2.1 Closer relationship between services and digital goods

productivity and the problems of measuring the effects of digital goods production is given in Sect. 2.4.

2.2.4.2 Software on the Borderline Between Goods and Services

The link between IT services and digital goods is strong: the (automatic) translation of text via a web service is an IT service whereas the stored result (e.g. an e-book) might be a digital good.[9] Contrary to digital goods, IT services cannot be stored. And, as opposed to digital goods, the utility of an IT service is often bound to a special time or customer. Economically speaking, it is easier to exploit scale effects by producing digital goods than by offering IT services.

Despite this definition, some products such as software can be classified as a service as well as a good. The value added by burning the code of an individually developed commercial software system onto a CD is negligible compared to the development costs of the software. From the economic point of view, the prevailing characteristic of the product has to be taken into consideration.

Anyhow, the links between services and digital goods are getting closer. Vickery and Wunsch-Vincent (2005) point out that the digital delivery of business services is growing, but the suitability 'is mainly affected by the centrality of information exchange, level of standardisation, complexity of task, nature of the problem and the knowledge involved, and the context of delivery'. On the other hand, digital goods (especially software) are increasingly bundled with additional service components such as automatic updates, consulting, etc. The bundling strategy can be interpreted as a form of versioning (see Sect. 3.1.4).

The classification of software as a good or service also is an unsolved issue with respect to the adaptability of WTO's trade regimes GATT (for goods) and GATS (for services – see also Sect. 9.3.3). According to Wunsch-Vincent (2005b), most WTO members agree that the majority of electronically-traded services should be governed by GATS, but they do not agree about the classification of digital products which are delivered on a physical medium like CDs or DVDs.

Statisticians solve the classification problem by introducing extensive rules for definitions. According to the Extended Balance of Payments Services (EBOPS)

[9] The SNA 1993 claims that the outputs of the service industries 'are often stored on physical objects (paper, tapes, disks, etc.) that can be traded like ordinary goods. Whether characterized as goods or services, these products possess the essential characteristic of being produced by one unit and supplied to another, thus making possible division of labour and the emergence of markets (UN (2002)).

classification, 'hardware and software consultancy and implementation services, maintenance and repair...design and programming of systems ready to use (including web page development and design)...development, production, supply and documentation of customized software, including operating systems made on order for specific users...web page hosting services and computer facilities management' are included as computer services. But on the other hand, all 'packaged (non-customized) software' is classified as goods (UN (2002)).

Following its double-sided character, software is mentioned as a digital good and an IT Service in this research, emphasizing the high level of standardization of B2C and standard operating systems software on the one hand, and the high degree of individuality of many B2B software tools on the other.

2.2.4.3 Open-Source Software

Open-source software is a special case of an internationalization pattern, which is not in the focus of the theoretical models of Part II: The internationalization by cooperation between developers. Open-source software can be downloaded, copied and used without paying fees, and the user can review the source code.

The website http://www.sourceforge.net is the most important website for open-source software. It listed 75,000 open-source projects at the beginning of February 2006, which are subject to 60 different licences. In the late 1990s, the Open Source Initiative introduced the umbrella term 'open source' to describe the sets of different types of free licences (Stenborg (2004)). The most important licences are the General Public License (GPL, 51,000 projects on Sourceforge.net), the Lesser General Public License (LGPL, 8,469 projects), and the Berkeley Software Distribution (BSD) License (5,391 projects).[10]

GPL ensures that all code released under it will remain free software. The open-source community calls this licence 'viral' or 'copyleft' (instead of 'copyright') (Henkel (2004)). Key elements of GPL are (Forge (2004)):

- Everyone is allowed to copy, distribute and modify the software; all modifications must be placed under GPL as well;
- Once the software is placed under GPL, it cannot be revoked;
- GPL software comes without warranty.

At the moment, a third updated version of GPL is being discussed in the international open-source community (FSF (2006)). One key issue is the relationship between GPL software and Digital Rights Management (DRM) systems.

Unlike the General Public License, under the Berkeley Software Distribution licence it is allowed to integrate proprietary code into software and to redistribute the work commercially. LGPL renounces the 'viral effect' of GPL, which

[10] Unfortunately, a breakdown of projects regarding licences and languages is not available any more on Sourceforge.net since September 2007. The total number of projects registered amounts to 106,498.

means combinations of LGPL, and commercial software can also be distributed commercially.

Lerner and Tirole (2002) explain the choice of licences with a small microeconomic model. Analysing the Sourceforge.net 2002 data, they found that highly restrictive licences are less common for more mature projects, projects running on proprietary operating systems, and projects dedicated to software developers. However, restrictive licences are more common for applications addressing end-users and system administrators. Also exploring the Sourceforge.net data, Comino et al. (2005) find that projects distributed under highly restrictive licences such as GPL have a significantly lower probability of reaching an advanced stage of development. They also found that the size of the community of developers has a positive but decreasing effect on the chances of progress, maybe due to coordination problems of large communities. According to Lerner and Tirole (2002), the following determinants are crucial when the developers are choosing which licence should be used when registering the project initially:

- Motivations of the developers;
- Project characteristics (environment, size of code base, intended audience);
- General considerations, such as the possible interactions with other projects, the possibility that the code is 'hijacked' (when a commercial software vendor adds some proprietary code and makes the whole project private, it is de facto privatization of code), impact of software patents, impact on the incentive to produce complementary software, familiarity of the open-source community with the licence, and 'forking' (the internal threat of incompatibility when competing developer groups move in different directions).

Economists try to explain the open-source movement with traditional microeconomic models. The developers' time spent is not rewarded by money but by:

- Gaining peer recognition, also called ego gratification incentive by Lerner and Tirole (2000);
- Career concern incentive, e.g. signalling skills to (future) employers;
- Learning incentives;
- Improving performance in paid work by using open-source components;
- Having fun choosing a 'cool' open-source software (it is more fun than working on a routine task set by an employer according to Lerner and Tirole (2004)).

Levine (2004) points out that these non-monetary benefits are only one aspect of open-source development and the instances of collective, coordinated online innovation also have to be explored.

Open-source developers can be freelancers who may have intrinsic motives, but companies also support open-source programming. Hawkins (2004) identifies the reasons for a profit-orientated company to support open source and to declare its software as open source:

- Gaining support contracts (Red Hat's business model);
- Reducing costs of development and testing;

- Bundling, especially for hardware manufacturers (e.g. IBM), and creating new demand for the hardware (e.g. SUN): in an empirical study, Fosfuri et al. (2005) found that companies with a strong market presence in hardware (downstream) and a strong position in proprietary software technology (upstream) are more likely to favour open-source software;
- Selling commercial versions of software with additional features (e.g. VA Software sells a commercial version of Sourceforge.net as a software development management tool);
- Inducing additional revenues when operating a portal for this software through advertisements, etc.;
- Developed software is predominantly used internally;
- Exploiting innovations (Creber (2004));
- Emotional reasons: fighting against the dominance of Microsoft and its 'adhesion contracts' (Haering (2004)).

On the other hand, companies use open-source software not only because it is free to download. Forge (2004) analyses the economics of security, reliability, compatibility, quality control and performance of open source, which explains the wide use of open source in professional environments. According to Netcraft.Com (2007), 53% of all active websites use Apache, an open-source webserver distributed under the Apache licence, whereas 33% use a webserver from Microsoft (as in August 2007).

The international character of the open-source community is obvious. Although all source code is written in English, the front end of a computer programme (what the user can see) can be translated from English to other languages. According to Sourceforge.net data, open-source projects have been available in several European languages by February 2006 as shown in the following overview. Russian was not listed.

Czech (324 projects)	Danish (227 projects)	Dutch (879 projects)
Esperanto (57 projects)	Estonian (34 projects)	Finnish (172 projects)
French (3,815 projects)	German (5,820 projects)	Greek (148 projects)
Hungarian (284 projects)	Italian (1,347 projects)	Latvian (37 projects)
Lithuanian (51 projects)	Maltese (2 projects)	Polish (643 projects)
Portuguese (366 projects)	Slovene (69 projects)	Spanish (2,845 projects)
Swedish (403 projects)		

Due to automatic translation programs, these numbers do not say whether developers from the countries mentioned are involved. Most popular open-source projects are translated into several languages. For example, eMule, a popular file-sharing program developed under the GPL, which has been downloaded 356 million times since registration of the project (May 2002) till September 2007. According to Soureforge.net statistics, it is available in 35 different languages. Despite all caution in interpreting the numbers, these clearly show the internationality of open-source software.

2.3 Recent Developments

2.3.1 Networks

2.3.1.1 Overview

Today, the Internet as a 'network of networks' (Ida and Ueda (2004)) dominates the scene. For information on the layer structure of the Internet, see Sect. 2.1.2. To measure the spread of the Internet, several indicators can be used:

- Demand side: time or money spent, number of subscribers, etc.
- Supply side: number of hosts, web servers. This can be split further, e.g. into the number of secure servers, IPv4 addresses, etc. (see OECD (2007f)).
- Domain names: number of domain names registered. It can be differentiated further into ccTLDs (country-specific domains such as .de), and gTLDs (.biz, .org, etc.).

Figure 2.2 shows the number of subscribers to the Internet in OECD countries, including mobile connections. According to the OECD definition, the subscribers are shown as 'broadband' subscribers when the downstream rate is greater than 256 kbit/s. Till June 2006, the number of broadband subscribers had grown to 178 million (OECD (2007f)). A classification akin to that of Internet subscribers is that of Internet users. The number of Internet users is higher than the number of Internet subscribers since two or more consumers can collectively use Internet access.

2.3.1.2 Peering and Transit

'Peering' is an agreement between Internet Service Providers (ISPs) to carry traffic for each other. It does not include carrying traffic to third parties. Historically, there is no explicit exchange of money between parties ('bill-and-keep'). The costs of the

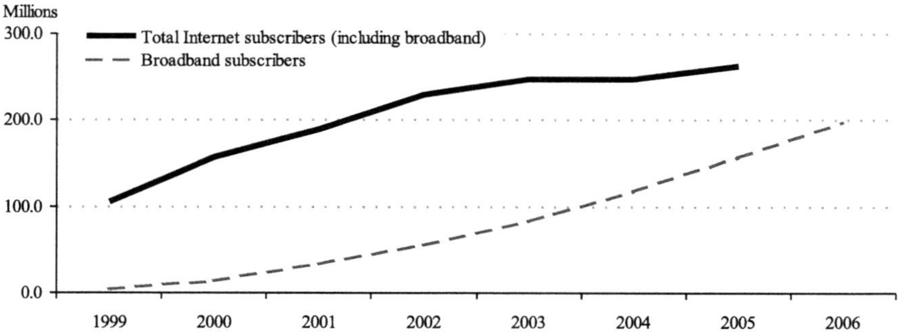

Fig. 2.2 Internet subscribers in OECD-countries (in millions), Source: OECD key ICT indicators

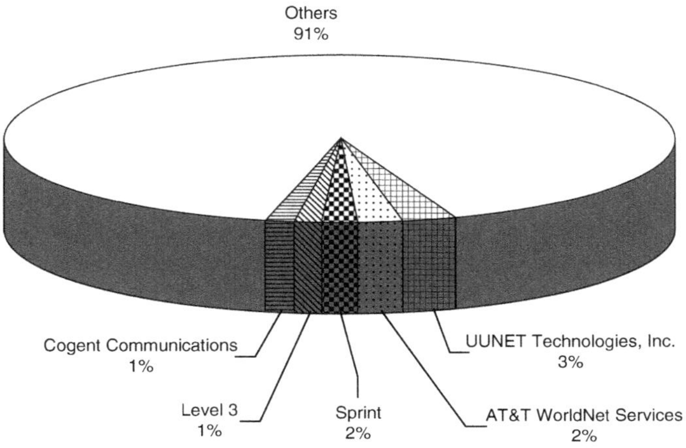

Fig. 2.3 Top five networks by number of peers in August 2006; Data source: OECD (2007f)

interconnect itself are shared. 'Paid peering' is used only if the parties do not receive 'roughly equal value' (ITU (2006b)).

Referring to data from http://www.fixedorbit.com, OECD (2007f) reports 95,000 peerings in August 2006, up from 79,000 peerings in September 2004. The five networks with the highest number of peers operate less than 10% of all peers (see Fig. 2.3). These data reflect the decentralized organization of the Internet network worldwide where various routes (paths) can be followed (see Fig. 9.1). Despite the decentralized peering structure, the access to the customer (last mile) might be a question of regulation, especially when a former incumbent is involved (see Sects. 9.1.2 and 9.1.3 also).

In contrast to peering, a 'transit agreement' between ISPs has a specific price as it is the obligation of the transit provider to carry traffic on behalf of another ISP also to third parties (ITU (2006b)).

2.3.1.3 Price Development

Declining prices of computing power (see Sect. 2.3.4) also influence hardware such as routers and switches, which are necessary to operate a data network. A study presented by ITU (2006a) shows about the global, average development of prices (downward sloping) and speed of connectivity (upward sloping curve) for end-user connections.

The price decline is also remarkable at the wholesale level. OECD (2006d) reports that the average price for a 155 Megabit per second (Mbps) transit connection was more than USD 500 per Mbps in London in 2001. By the second quarter of 2005, the equivalent transit connection could be purchased for around USD 40 per Mbps.

2.3.1.4 Recent Innovations

Recent innovations in the network segment are based on broadband and mobile access which bundle services. Some new acronyms particularly dominate business talk.

- In general: **NGN – Next Generation Networks:** according to ITU, NGNs are a 'packet-based network able to provide telecommunication services and able to make use of multiple broadband...and in which service-related functions are independent from underlying transport-related technologies'. In short, providers migrate from analogue telephone (PSTN), mobile and IP-network infrastructures to unified next-generation networks (Ponder (2006)). NGNs usually offer broadband access to their users. (For the effects of NGNs on economic policy, see Sect. 9.1.4.)
- Mobile access: different types of waves (frequencies) and encoding technologies (modulation) are used which determine speed and availability of access. The most important acronyms are (see ITU (2006a), OECD (2007f), and OECD (2007d) for more information):
  - Long distance: **UMTS – Universal Mobile Telecommunications System**: a standard for third generation (3G) mobile phones. The abbreviation UMTS is used in Europe; internationally, the transmission protocol is called W-CDMA (Wideband Code Division Multiple Access).
  - Middle distance: **WiMax – Worldwide Interoperability for Micro-Wave Access:** WiMAX is designed to transmit data up to 70 Mbit/s over a maximum range of 50 km.
  - Short distance: **WLan – Wireless Local Area Network:** hotspot access at short distances. The most popular WLAN standard is called Wi-Fi (wireless fidelity). A new standard which is able to transmit up to 540 Mbit/s was presented in 2007.
- Fixed line networks:
  - Copper cable: **xDSL – Digital Subscriber Line:** the x can be replaced by S (single-pair high-speed), A (asymmetric) or V (very high-speed digital subscriber line). xDSL describes technologies to transmit data at high speed (up to 100 Mbit/s downstream and upstream with 'VDSL2') using old-fashioned copper cables (ITU (2006a)).
  - TV cable: TV cable providers additionally offer telephone and Internet access.
  - Fibre: signals are transmitted optically, which allows very fast data traffic. Fibre cables are especially used between data centres.
- Quad-play: Bundles which include a combination of television, Internet, telephone (fixed line), and mobile services (see also Sect. 9.4)

2.3.2 *IT Services*

2.3.2.1 Overview

Due to the lack of reliable statistical data about the global service sector, the volume of the IT services has to be estimated. An estimation is offered by Pierre Audoin Consultants (PAC), who estimated the volume of the worldwide IT service market to be Euro 424 billion in 2006. Included are outsourcing operations worth Euro 161 billion (IDATE (2007)).

In recent years, a driving force for the IT services sector has been the establishment of IT links between retailers, intermediate, and final goods producers. For a country- and sector-specific overview of the share of companies which (sometimes) use the Internet to purchase or sell products, see Appendix 12.2.

Another indication for the growth of IT services is the number of worldwide websites. Statistics from Netcraft.Com (2007) show that the number of all registered website addresses (hostnames) and the number of active web pages have exponentially grown in the last decade.

2.3.2.2 Outsourcing and Offshoring

Thanks to balance of payments statistics, there are more appropriate data available about the trade in IT services. According to IMF data, world exports of computer and information services had a volume of USD 71.5 billion in 2003 (UNCTAD (2006)). Figure 2.4 shows the growth path of computer and information services compared to 'ICT-enabled services'. This category also encompasses insurance services, financial services, royalties, etc. The total trade volume of ICT-enabled services had been USD 829.6 billions in 2003.

Trade in services is also called outsourcing or offshoring. I will pay special attention to them in this section. Outsourcing can be achieved domestically or

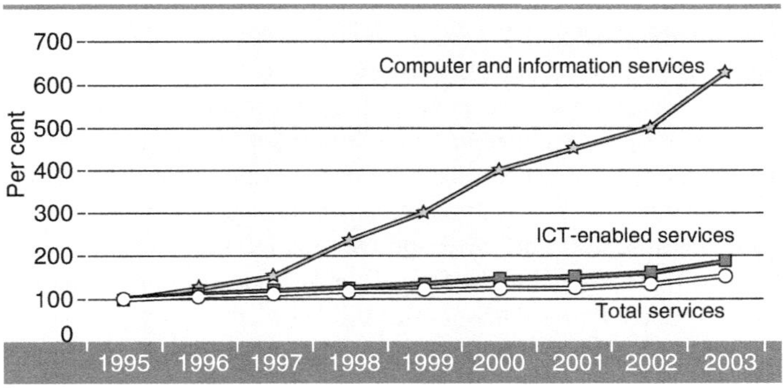

Fig. 2.4 Growth of world services exports; Source: UNCTAD (2006), based on IMF BOP data

Table 2.9 Classification of trade in services; Source: EC (2005a), p. 92

	Ownership of activity: internal to the firm	Ownership of activity: external to the firm
Location of activity: home	Domestic in-house production (Firms produces its products domestically without any outside contracts)	Domestic outsourcing (Firms uses inputs supplied by another domestically-based company)
Location of activity: overseas	Offshoring (Firm uses inputs supplied by its foreign-based affiliates)	International outsourcing (Firm uses inputs supplied by an unaffiliated foreign-based company)

internationally, called offshoring. EC (2005a) defines offshoring as the use of inputs from a foreign-based affiliate, e.g. internal to the firm. For an overview of the EU definition, see Table 2.9.

Other authors differentiate between internal (or captive) offshoring, and offshore outsourcing (see OECD (2004b), for example). However, these distinctions only become relevant when analysing the behaviour of multinational companies (MNCs). Boes (2005) points out that pure IT service companies prefer internal offshoring. However, from the economic point of view, both internal offshoring and international outsourcing imply the import of intermediates (services) as is stressed in the fragmentation literature. International IT outsourcing implies the same advantages and risks as domestic IT outsourcing, but the amplitude may be larger due to higher differences in labour costs: according to Allweyer et al. (2004), a high-quality software developer earns Euro 120–200 a day in India, whereas in Germany the daily rate is Euro 600–1,000. Cost-orientated international outsourcing may lead to the effect that the value added measured at company level rises, whereas the value added at country level shrinks. Other reasons for offshoring occur when international IT resources are needed for international expansion. This may concern the availability of language capabilities in special time zones (e.g. for a call centre) or IT support for foreign affiliates. A more comprehensive look at the motives for outsourcing and offshoring is given in Sect. 3.3.1.

The empirical picture is not clear overall. Huws et al. (2004) argue that the international outsourcing business is now changing from an experimental to a consolidation phase. As products become more complex, value chains become longer. In the ICT producing sector, there has been a strong trend for the international outsourcing of low-skilled activities to be increasingly supplemented by higher-skilled activities, such as research and development (R&D). On a macroeconomic level, Amiti and Wei (2005) argue that the fear of service outsourcing dramatically reducing job growth in the advanced economies has been greatly exaggerated. They find that most developed countries are not generally more outsourcing-intensive than many developing countries. Using data of the IMF, the authors rank the countries as follows. The biggest outsourcers of business services are the US (USD

41 billion in 2002), Germany (39), Japan (25), and the Netherlands (21). The four biggest outsourers of computer and information services are Germany (6.1), the UK (2.6), Japan (2.1), and the Netherlands (1.6). On the other hand, the four biggest insourcers of business services are the US (59), the UK (38), Germany (28), and France (21). The four biggest insourcers of computer and information services are Ireland (10.4), the UK (5.7), the US (5.4), and Germany (5.2). For further estimations of the volume of international outsourcing, see also OECD (2004b), although the report reveals enormous statistical discrepancies, exemplary for the case of India. Kirkegaard (2004) explores the case of the US-India trade balance in detail and describes several reasons for the discrepancies. Nevertheless, the OECD (2004b) report emphasizes that cultural similarities are important for international outsourcing, noting that international outsourcing has been dominated by American and British companies which offshore operations to India, Canada, and Ireland. Another OECD report (Welsum and Vickery (2005)) directly confesses that 'little is known about the impact and extent of international outsourcing' by official statistics. But it states that 'anecdotal evidence suggests that the international sourcing of IT and ICT-enabled services is growing rapidly', and estimates that 20% of total employment in the EU15 could potentially be affected by the international sourcing of service activities. Mann (2005) points out that outsourcing not only influences the number of IT jobs, but also the mix of IT skills demanded – 'with a greater emphasis on integrating imported components, and analyzing, designing, and implementing IT products'.

WTO (2005f) emphasizes the rapid growth of world exports of computer and information services, which was 31% (cumulative) over the 2002–2003 period. Ireland and India were identified as the two major exporters of computer and information services in 2003. According to EITO (2005), offshoring services are defined as services sold in Western Europe but delivered from lower-cost countries (such as the Czech Republic, Poland, Slovakia, Russia, India, China, and South America). This represents some 1.1% of total IT service spending in Western Europe. It is further estimated that this ratio will rise to 1.8% in 2008.

2.3.2.3 Recent Innovations: Service-Oriented Architectures

The future development of international production sharing may also be influenced by technological improvements. Even if companies accelerate international outsourcing efforts, Carr (2005) predicts that four technical factors drive the centralization of IT services:

- Virtualization, which erases the properties between computing platforms from the user's point of view.
- Grid computing: a large number of hardware components which act as one device.
- WebServices (see below).
- Fibre optics networks, for quick access to large data volumes.

The way data exchange between companies is organized is pivotal. For a long time, Electronic Data Interchange (EDI) has dominated as a general concept. Using EDI, two companies exchanging data had to agree on a standard, which had been 'in many cases a costly negotiation, contraction, and software adaption process' (Kuropka and Weske (2007)). The key idea of the more recent concept of 'service-orientated computing' is to provide well-described services, which can be invoked by a published interface. It can be performed on demand, particularly also by demand of another computer. To make this applicable, programmers have to take into account service-oriented architectures (SOA). Three types of participants are crucial in service-orientated applications (Kuropka and Weske (2007)).

- Service providers, which offer the services and own the necessary resources.
- Service brokers, which have the role of 'yellow pages'.
- Service requestors, which are the customers.

Related to SOA is the key word of 'WebServices' (WS). WS are a 'manifestation' (Kuropka and Weske (2007)) of the SOA with explicit use of the Internet for data exchange. WS and SOA emphasize distributed and network-based approaches of computing in which the network itself is the source of computing power. So Mcafee (2005) describes WebServices as replacing interactions between humans and applications, by interactions between applications only. The concepts are crucial to 'enable enterprises to collaborate in shared business processes' (UNCTAD (2006)). Outsourcing and offshoring (see Sect. 2.3.2) benefit from SOA technologies.

Open standards are crucial for service orientated architectures to work. But Kuropka and Weske (2007) point out a semantic problem: making the meanings of words clear in programme codes (e.g. parameters, objects, etc.) could be a problem. In earlier times, the software vendor could be asked or the documentation checked. With computers negotiating automatically, there is no easy way to handle alternatives.

The market volume of WS is estimated by several consulting agencies. According to UNCTAD (2006), consulting agency IDC forecasts WS expenditures to USD 14.9 billion in 2009, Radicati Group estimates USD 6.2 billion by 2008, and Gartner predicts the 'market for IT services related to WS' would amount to USD 261 billion worldwide by 2008.

2.3.3 *Digital Goods*

2.3.3.1 Overview

Data about digital goods are not collected by national statistics. In contrast to some well-documented ICT indicators, the global trade volume of digital goods is unknown. Nor can the volume of national e-commerce trade, which is based on a broader concept than digital goods, be determined precisely.

Table 2.10 Distance trading in Germany (2007 estimates)

Category	Volume in billion Euro	Percent	Source
GDP 2007 (estimated)	2,429		SVR (2007) and own calc. based on price forecasts
Consumer spending	1,347	100	SVR (2007) and own calc.
..Distance trading ('Versandhandel' – only physical goods)	27.6	2.05	BVH (2007)
..Online shopping (goods, services and digital goods)	16.8	1.25	BVH (2007)
..Physical goods bought online (incl. CDs, DVDs, etc.)	10.9	0.81	BVH (2007)
..Services (travel, etc.) bought online	5.0	0.37	BVH (2007)
..Digital goods (mp3s, ring tones, games and software)	0.6	0.04	BVH (2007)
Graphical online advertisements Jan–Sep 2007	0.6	/	IDATE (2007)

Nevertheless, the significance of digital goods can be guessed when looking at survey data and/or studies which focus on sector-specific issues (see Sect. 2.3.3 for the music industry, for example). A comprehensive study about 'distance trade' in Germany has been done by Bundesverband des Versandhandels (bvh). The study is based on a survey of 24,000 consumers and also estimates the volume of digital goods sold in Germany (see Table 2.10).

Services bought online have a higher volume in euros than digital goods because airplane and railway tickets have a higher price than, for example, music files. Measured in monetary units the significance of digital goods for an economy is small, but when asked what had been their last online transaction, about 16.1% of the consumers buying services and digital goods online reported that they had bought a music file, 13.0% an airplane ticket, and 10.6% a ticket for a concert.

The estimated figure of Euro 600 million for sales of digital goods in line eight of Table 2.10 is related to the B2C market only, wholesales are missing. Therefore the figure underestimates the total digital market volume (see Sect. 2.3.4 also). For example, the volume of graphical online advertisements, a typical B2B application (such as banners, pop-ups, and videos) totaled Euro 627 million in the first three quarters of 2007 (Bitkom (2007)).[11] Furthermore, parts of the TV services market, which had a total volume of Euro 11 billion in Germany in 2006 (IDATE (2007)), may be added.

[11] A rise of 68% compared to Jan–Sep 2006.

Another figure is estimated for the US: the US Census Bureau mentions that e-commerce sales[12] have attributed 3.3% of total retail sales in second quarter 2007 (Uscensusbureau (2007b)). Unfortunately, digital goods are not stated separately.

2.3.3.2 Music Industry

Music files such as .mp3 are a good example of digital goods as defined in Sect. 2.2.3. Worldwide total music sales summed up to USD 7,853 million in the first half of 2007 according to IFPI (2007b). On average, 17% of this was distributed online.[13] Online distribution is most used in Korea (63% of total), and the US (29%). In Germany, online sales contributed 7% to total music sales in January–June 2007. Korea is the only market where online sales have overtaken physical delivery of music. One reason is the early introduction of a flat rate by South Korean SK Telcom's MelOn: for a monthly flat rate of around USD 5, telecom customers can stream and download DRM (digital rights management)-protected music. Another reason is the wide spread of mobiles, which can be used as a music device too (IFPI (2007a)).

Historically, the use of the carrier medium of music shows distinctive life cycles, so the introduction of CDs had displaced vinyl and cassettes. Note that music files on CD and DVD are already digital. Digitalization of music combined with widespread access of music consumers to the Internet has led to (Bourreau (2005)):

- Separation of music from the physical carrier and
- New sound formats.

Separation from the physical carrier influences the music industry. A threat is that users exchange music directly without loss of quality. Peitz and Waelbroeck (2005) show that CD sales have shrunk sharply since 2000, and explain a part of it with the introduction of Napster, a so-called Peer-to-Peer(P2P) service which supports direct exchange.[14] The music industry has reacted with the introduction of DRM (Digital Rights Management) Systems (see also Sect. 9.3.2), and took legal action against companies and users providing or using P2P services.

Also, new business models had to be introduced. Marginal costs have changed through separation from the physical carrier. On a CD, several songs are bundled

[12] The US Census Bureau defines e-commerce sales as 'sales of goods and services where an order is placed by the buyer or price and terms of sale are negotiated over an Internet, extranet, Electronic Data Interchange (EDI) network, electronic mail, or other online system. Payment may or may not be made online.' Source: Uscensusbureau (2007b).

[13] According to IFPI, an association of the recording industry, this includes single-track downloads, album downloads, music video online downloads, streams, master recording ringtones, full-track audio download to mobile, ringback tunes, music video downloads to mobile and subscription income. Excluded from these figures are ringtones and non-music content. Growth rates are calculated on fixed USD values using IMF average monthly exchange rates.

[14] Oberholzer-Gee and Strumpf (2007) argue with empirical data that the opposite should be true. But their paper is criticized vehemently for scientific shortcomings; see Haering (2008).

in a 40–80 min programme, while with selling music online, unbundling prevails: offering a single song for USD 0.99 has been a popular strategy (Peitz and Waelbroeck (2005)). Furthermore, new sound formats, which allow encoding of a music file with different size and quality, have been introduced. The differences in quality may lead to price-discriminating strategies.

A recent trend is that online music providers interpret their business model as two-sided markets (see Sect. 3.2) and try to 'collect' customers, so they have relinquished DRM technology. For example, iTunes, a popular service of PC distributer Apple, offered two million DRM-free songs for 99 cents by October 2007.[15]

2.3.3.3 Recent Innovations

The trend of digitalization transforms classical analogue products into digital ones. This concerns TV and telecom services particularly. For Voice over IP (VoIP), see Sect. 2.3.4.

Other innovations, where digital goods are usually bundled with specific services, are related to social or marketing aspects. The most important phrases are user-generated content (such as www.youtube.com), social networks (www.myspace.com), virtual worlds (www.secondlife.com).[16] Participative Web or Web2.0 are general terms for these new developments (see OECD (2007g)). The business model of social networks is best described by the theory of two-sided markets, which is presented in Sect. 3.2 and linked to macroeconomic variables in Sect. 6.3.

2.3.4 Related Topics

2.3.4.1 Computer Power and Price Measurement

Digitalization cannot be thought of without computers. Networks, IT services and digital goods are affected by the technological progress which leads to rising performance and declining prices of computer power. To gain a long-term impression, Nordhaus (2007) defines 'million standardized operations per second' (MSOPS) as an indicator of computer performance. Approximately 1 MSOPS is equivalent to a machine which can add 20 million 32-bit integer numbers in one second. In Fig. 2.5, the costs per MSOPS are shown deflated with a consumer price index, base year 1998. The different technologies describe the core technology of the computing engine.

---

[15] Source: Apple Press notice: 'iTunes Plus Now Offers Over Two Million Tracks at Just 99 Cents', http://www.apple.com/pr/library/2007/10/17itunes.html.

[16] For example, the Korean company Cyworld operates a virtual world and reported that it had a revenue of USD 77 million (77.5% of total revenues) from selling 'goods' which can be used by the avatars in 2005 (Hillenbrand (2007)).

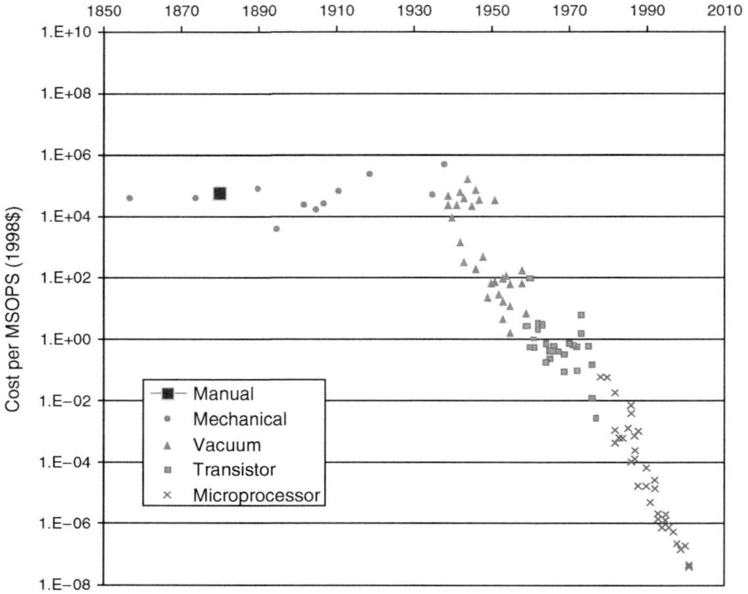

Fig. 2.5 Costs of computing power for different technologies; Source: Nordhaus (2007)

Nordhaus (2007) reckons that the cost of calculation as measured by MSOPS has fallen by a factor of 1,300,000,000,000 relative to consumer prices and by a factor of 5,100,000,000,000 relative to the cost of labour since 1900. The average increase of computing power was 79% p.a. in the period 1990–2000.

Another indication of accelerating computer power is Moores Law, named after Gordon Moore, a co-founder of chip-making company Intel. In 1965, Moore pointed out that the number of transistors per square inch on integrated circuits doubled every year. Reinvestigating the topic, he forecasted a doubling every 18 months. From today's point of view, Moore's forecast has proven to be quite accurate (Barnett et al. (2003)).

The estimations not only offer a description of the past, but can be used for prediction. Assuming that the current growth rate of computing capabilities will continue, Nordhaus (2007) reckons that a petaflop-computer will cost USD 2,000 in the year 2025 and USD 1 in the year 2035. 'Petaflop' machines execute one billion MSOPS, which is equal to the computational capacity of a human brain. He concludes: 'If nonhuman capital with human capabilities costing virtually nothing is indeed a serious possibility in the next half century, then the organization of economic and social activity in such a world should be high on the research agenda today' (Nordhaus (2007), p. 31).

The enhancements of computing power are pivotal when price indices of computers have to be calculated. Nordhaus (2007) has used a standardized measure and could calculate the price decline per MSOPS. But in computer retail shops, 1 MSOPS can not be bought separately, as computers are bundled systems with different quality and speed. Furthermore, computers have a life cycle, so that the basket, which is used by the statistic agencies to calculate a computer price index, changes permanently.

To deal with theses issues, statisticians have developed a set of quality adjustment procedures. A detailed overview is given by Triplett (2004). In short, the procedures can be differentiated:

- Quality adjustments in conventional price index methodologies: For example, a method of this category is the 'overlapping link' method. Assume that in period t a new computer n is introduced. The higher price reflects the improved quality compared to older model m. Then the prices for periods $t-1$, t, $t+1$ are given by $p_{m,t-1}$, $p_{m,t}$, $p_{n*,t-1}$, with p_{n*} prices adjusted by factor $p_{m,t}/p_{n,t}$ for all future periods.
- Hedonic price measurement: To calculate a hedonic price index, a hedonic quality adjustment in two steps is worked out:

  1. A regression to calculate the 'hedonic function'. For example:

  $$P_i = c + \beta_1 GB_i + \beta_2 RAM_i + ... \tag{2.1}$$

     with P_i for the price of computer i, c constant, GB volume of hard drive in gigabytes of i, RAM access memory of i, and more characteristics of computer model i.
  2. The estimated coefficients β_1, β_2,.., β_n of the hedonic function are then used to calculate the hedonic quality adjustment A.

  $$A = \beta_1 \frac{GB_n}{GB_m} + \beta_2 \frac{RAM_n}{RAM_m} + ... \tag{2.2}$$

     When the new computer model n is introduced in period (t+1), $P^*_{n,t} = AP_{m,t}$ with $P_{m,t}$ for the price of older model m in period t gives a theoretical price of model n in t. Accordingly, the computer price index for (t+1) can be calculated considering the price change of n from t to t+1, $P_{n,t+1}/P^*_{n,t}$.

  This second step is the method used most often to calculate hedonic price indices. Other methods (time dummy variable method, characteristics price index method, hedonic price imputation method) are explained by Triplett (2004).

The US Bureau of Labor Statistics (BLS) uses the hedonic method described above. According to Nordhaus (2007), BLS uses computer components to build the hedonic indices, whereas Nordhaus uses performance measured in MPOPS. This has the effect that the nominal price decline rates calculated by Nordhaus (41% p.a. in 1990–2000) are higher than the rates calculated by BLS (13% p.a.

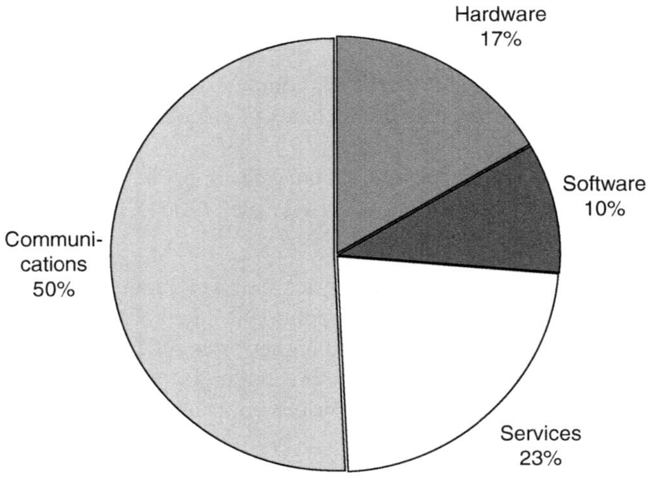

Fig. 2.6 World ICT market expenditures in 2005; Data source: OECD (2007f), based on WITSA figures

in 1990–2000). Another reason is that Nordhaus refers to computers only, whereas BLS also includes peripheral equipment.

2.3.4.2 ICT

The concept of Information and Communication Technology (ICT) is related to the network, IT services, and digital goods approach presented here (see Sect. 2.2.4). Figure 2.6 shows that communications dominates the ICT sector. Total ICT expenditures were USD 2,963 billion in 2005 (OECD (2007f) based on figures of the World Information Technology and Services Alliance (WITSA)).

The compound annual growth rate (CAGR) of worldwide ICT expenditures (current prices) were 5.6% in the period 2000–2005. As ICT subsector price indices have fallen the rate of ICT investments in real terms is higher.

Comparing 25 industries across seven countries, Inklaar and Timmer (2007) found that almost all industries show positive ICT capital deepening in the period 1987–2003. The links between ICT capital deepening and productivity growth are described in Fig. 2.10. Moreover, there are a number of studies which offer anecdotal evidence for the productivity-enhancing use of ICT in specific sectors. For example, UNCTAD (2006) reports about ICT use in the oil sector; several sectors (food and beverages, footwear, pulp and paper, ICT manufacturing, consumer electronics, shipbuilding, construction, tourism, telecommunications and the hospital activities industry) are covered in EC (2006a).

The data from OECD (2007f) can also be used to analyse the correlation between hardware, software, services, and communications expenditures. Table 2.11 shows

Table 2.11 Correlation matrix of global ICT expenditures: 2000–2005 (own calculation)

	Communications	Hardware	Services	Software
Communications	1	0.85	0.97	0.97
Hardware	0.85	1	0.77	0.74
Services	0.97	0.77	1	0.99
Software	0.97	0.74	0.99	1

that there has been a very strong correlation between global expenditures for software, communications in ICT services in the period 2000–2005, and a high correlation of these segments to hardware. Economically, this result may be interpreted as the several ICT goods and services being complementary.

2.3.4.3 Voice over IP

Voice over Internet Protocol (VoIP) services are a further example of a category of products which range between digital goods and services. Modern network topology allows transfer of all kinds of digital data, which can be voice as well (VoIP). For a brief comparison with older PSTN technology, see Sect. 2.2.1.

VoIP is a digital good as it is aspatial, recombinant, and the VoIP service provider can expand its supply of 'VoIP over Internet' connections easily. But on the other hand, VoIP has service characteristics also since the voice connection is directly related to two (or more) partners and the telephone service call itself is not storable.[17]

From an industry supply-side perspective, three key groups offer VoIP services.

- **VoIP over Internet provider:** VoIP providers which offer connections between Internet users. Their service is to induce the call ('origination'). After that, the two users are connected directly.
- **Gateway provider:** They offer gateways between VoIP users and PSTN users. The terms of interconnection with traditional PSTN suppliers are crucial for this group of companies.
- **Traditional telephone and cable TV providers:** Telephone providers replace circuit-switched technology and use VoIP technology for their internal (long-distance) backbones. Cable TV providers bundle their telephone services with other services such as TV and Internet access ('Triple Play').

Slightly different groupings can be found at Bijl and Peitz (2006) and OECD (2007f). Monopolkommission (2006) differentiates according to the bundling of VoIP with other products: VoIP with own broadband or Internet access, VoIP with leased local loop or bitstream access, and unbundled VoIP (see also Sect. 9.4.2).

[17] Of course, the dialogue can be stored in an audio file. But this does not replace the phone service that enables two persons to speak to each other over a distance.

Pricing of VoiP differs from traditional PSTN pricing. With circuit-switched technology, every connection needs its own cable, whereas with digital packets, a cable can handle several connections and routing is more flexible. Presuming that Internet access is already available, there are nearly zero marginal costs for a voice connection between two users via VoIP over Internet. The provider does not need to provide server capacities because a direct connection between the two users is established.

Providers of the second group need a termination in the PSTN network. They are charged by the traditional telephone companies accordingly. The most prominent example of this group is Skype, a sister company of Ebay.[18] The third group uses VoIP to reduce costs. A precursor is BT, the British telephone incumbent.

Summarizing, VoIP leads to 'a strong downward pressure of prices for voice services' (OECD (2007f)) and the development of new services (see OECD (2006h) for examples). Apparently, the 'digital goods part' of VoIP characteristics also influences the pricing policy of traditional telecom providers. The regulatory aspects which are related to VoIP are discussed in Sect. 9.4.2.

2.3.4.4 E-Commerce

In the context of this research the trade of digital goods is a part of e-commerce, and e-commerce transactions are a driving force for the development of IT services and networks. According to OECD, an 'electronic transaction is the sale or purchase of goods or services...over computer-mediated networks. The goods and services are ordered over those networks, but the payment and the ultimate delivery of the good or service may be conducted on or offline' (OECD (2005)). A related definition of 'Internet transactions' excludes orders received or placed by telephone, facsimile, or conventional email. The US Census Bureau defines e-commerce sales as 'sales of goods and services where an order is placed by the buyer or price and terms of sale are negotiated over an Internet, extranet, Electronic Data Interchange (EDI) network, electronic mail, or other online system. Payment may or may not be made online' (Uscensusbureau (2007b)).

The above definitions are critical from two aspects. (1) Should goods be taken into account, where only the first contact was established online? This may be critical when dealing especially with worthy assets. (2) Are orders part of e-commerce when they are induced by email? National statistics agencies answer these questions variably (see also the first remark in Appendix 12.2).

The US Census Bureau reckons that in the manufacturing industry, USD 1,266 billion or 26.7% of all shipments are induced by e-commerce transactions, and in the merchant wholesale sector, 18.3% of total sales in 2005. The figures in the B2B sector are significantly higher than the share of e-commerce in the retail business.

[18] In its 2006 annual report, Ebay reported that 171 million users were registered by Skype as on 31 December 2006. The usage minutes of outgoing connections (from Skype to other providers) were 4.1 billion in 2006, which resulted in revenues of USD 178 million.

A reason is that proprietary Electronic Data Interchange (EDI) systems are more common in the B2B than the B2C segment (UScensusbureau (2007a)), although search engines and websites of companies, media and consumer organizations for search and information purposes play an important role preparing some consumers's decisions at 'offline' stores. In Appendix 12.2, the sales and purchases induced by Internet transactions are listed.

2.3.4.5 Digital Divide

The term 'Digital Divide' points to existing discrepancies in the possibility of accessing and using digital communication networks. Figure 2.2 shows the rising number of Internet subscribers worldwide. Figure 2.7 shows the growing equality of Internet use. If the rate of Internet usage were the same across all countries, the Lorenz curve would become the 45 degree diagonal line of equality. In 1997, 95% of Internet users lived in countries which represent 20% of the world's population. This skewed distribution can be interpreted as the digital divide between countries. The relationship became more equally distributed in the following years. The development may go on when Internet access through mobiles becomes available in all countries, since mobiles are the most equally distributed ICT good worldwide (ITU and UNCTAD (2007)).

The relationship between Internet access and income is striking. For other determinants of the digital divide, see Sect. 8.1.3 also. Figure 2.8 shows data of 30 OECD countries with GDP per capita and Internet users per 100 inhabitants in 2005.

Internet users are defined as individuals who used the Internet in the preceding 12 months. This number is higher than the number of Internet subscribers who pay for Internet access (UNCTAD (2005)). Here, the number of users is employed, since this data series is available for more countries. Running a regression for the

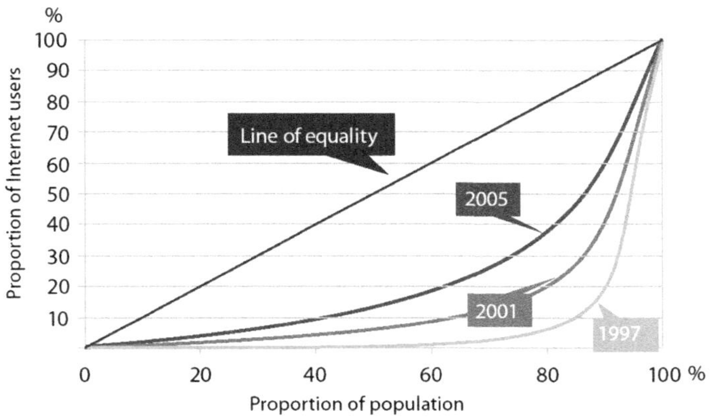

Fig. 2.7 Lorenz curve for worldwide internet users in 1997, 2001 and 2005; Source: ITU and UNCTAD (2007), p. 25

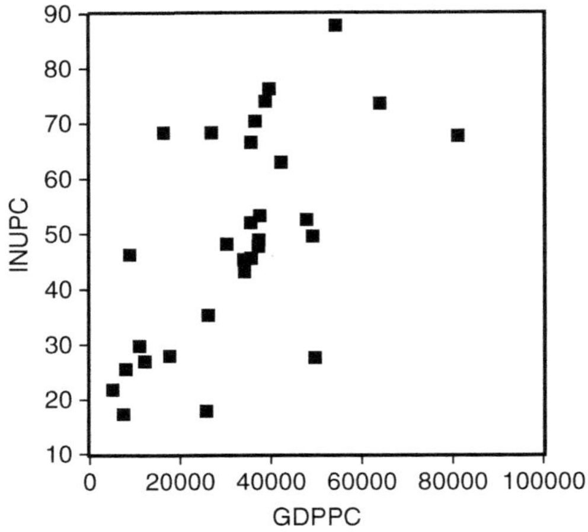

Fig. 2.8 GDP per capita (in USD per PPP) versus internet users per 100 in 30 OECD countries 2005; Data source: OECD and ITU databases

logarithm of GDP per capita against the logarithm of the share of Internet users shows a R^2 value of 0.45 and a value of the F-statistic of 22.98 (Prob. 0.000049). GDP and number of Internet users are interdependent.[19]

Figure 2.2 shows that a rising share of total access is established by broadband connections. The number of countries with broadband access rises. Macedonia had been the last European country without a broadband service, which was introduced in December 2003. Globally, in 170 countries, commercial broadband access with at least 256 kbit/s was offered in 2006, compared to 81 countries in 2002 (ITU and UNCTAD (2007)).

The digital divide is not neutral to technology. ITU (2006a) uses the Digital Opportunity Index (DOI, see Sect. 2.1.3) to measure the digital divide. The Gini score[20] amounts to 0.28 for total DOI, with highest value 0.96 for the criterion 'mobile broadband as percent of total mobile'. Grouping the countries according to income levels shows that the digital divide diminishes with the age of the technology. In Fig. 2.9, two groups of countries, high-income and middle/low-income, are considered; the total number of users is shown on the right-hand scale.

Overcoming the digital divide is a goal of economic policy which is generally accepted. Section 8.1.3 focuses on the political issues of the digital divide.

---

[19] The F-statistic indicates that with a 0.99 significance level, the number of Internet users has an impact on GDP. On the other hand, the results must not be overinterpreted: the problem of omitted variables is evident here. Therefore, the estimated β-coefficient is not worth mentioning here.

[20] Gini scores rank between 0 (perfect equality) and 1 (full inequality).

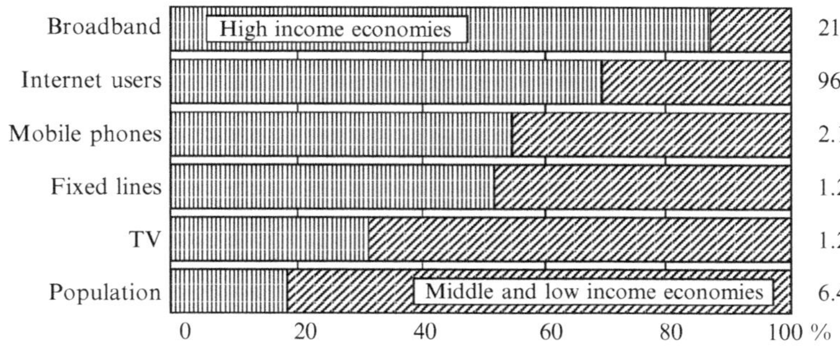

Fig. 2.9 Digital divide and worldwide users of technology in 2005; Data source: ITU (2006a)

2.4 The Impact for Productivity: Results from ICT Research

2.4.1 Introduction

An important link between digitalization and the macroeconomic environment is its influence on productivity. Measuring 'digitalization' as an input factor can be approximated by measuring the stock of information and communication technology (ICT). Therefore, ICT literature is at the heart of this chapter, while in Sect. 2.4.4 the slightly different requirements of measuring the productivity effects due to digitalization are discussed.

Considering ICT a decade ago, economists struggled with the so-called Solow Paradox: 'You can see the computer age everywhere but in the productivity statistics,'[21] That the IT sector was so small and the time lag of IT spillovers so long have been the main exculpations for this paradox. Meanwhile, the link between ICT and productivity has become evident at firm level, but also at sectoral or national levels. The next sections give an overview on related studies.

A basic idea of these studies is to consider ICT-producing and ICT-using sectors separately. In the ICT-producing sector (hardware, software producers, telecom suppliers, etc.), technical progress leads directly to a rise in Total Factor Productivity (TFP). The TFP concept is introduced in more detail in Chap. 2.4.2. In the ICT-using sectors, investments in ICT become part of the capital stock. There are two effects: ICT as general-purpose technology (Bresnahan and Trajtenberg (1995)) leads to the restructuring of the organization (see Sect. 2.1). Because this causes further investments, e.g. in education and training, the short-term effects on productivity of companies remains unclear. On the other hand, the labour productivity rises, which is a direct effect of capital-deepening. These effects are summarized in Fig. 2.10.

---

[21] Robert M. Solow, 'We'd Better Watch Out', *New York Times Book Review*, 12 July 1987, quoted by Howell (2004).

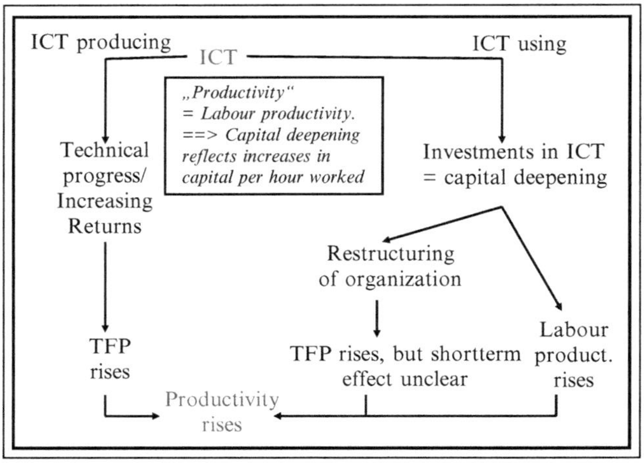

Fig. 2.10 Overview of contribution of ICT to productivity

2.4.2 Theoretical Foundations: Neoclassical Growth Assumptions

A general production function may be defined as:

$$Y = F(K_t, L_t, X_t, t) \tag{2.3}$$

with K_t for capital, L_t for labour, X_t a further input factor, and t as time.[22] Differentiating F to time and dividing by F = Y leads to the growth rate g in general:

$$g = \frac{1}{F(K_t, L_t, X_t, t)} \frac{dF(K_t, L_t, X_t, t)}{dt} = \frac{\dot{Y}}{Y}$$

This is equal to:

$$g = \frac{1}{Y}\left(\dot{Y} + \dot{K}K' + \dot{L}L' + \dot{X}X'\right)$$

with $K' = \partial F/\partial K$ and $\dot{K} = \partial K/\partial t$, for L and X respectively.

2.4.2.1 Neoclassical Case

For the standard neoclassical case of growth accounting, two assumptions are crucial:

- Perfect competition: The marginal products of the input factors equal the real factor prices. This also implies that the partial production elasticity of an input

[22] Sometimes technical progress is represented by a factor-augmenting parameter A, also described as Hicks-neutral: $Y = AF(K_t, L_t)$. But this notation does not change the methodology of calculating productivity growth $dlnA/dt$ as a residual.

factor equals the factor's share in income s with P_K, P_Y for the price indice of capital and domestic products.

$$s_K = \frac{P_K K}{P_Y Y} = \frac{\partial F}{\partial K}\frac{K}{F} \tag{2.4}$$

- Constant returns of scale: The partial production elasticities sum up to one.

Using these assumptions, the growth rate can be expressed as:

$$g = \frac{\dot{Y}}{Y} = s_K\frac{\dot{K}}{K} + s_L\frac{\dot{L}}{L} + s_X\frac{\dot{X}}{X} \tag{2.5}$$

The growth rate of Y is equal to the weighted sum of the growth rates of the input factors.

When the input factor X cannot be observed directly because it represents technology, organization, or an efficiency parameter in general, the difference between growth g and the measured growth explained by the observable input factors growth rates is called the Solow Residual (SR), also interpreted as Total Factor Productivity (TFP) or Multi-Factor Productivity (MFP) growth rates. With the two partial production elasticities of capital and labour summing up to one, this leads to:

$$SR = TFP_{GrowthRate} = \frac{\dot{Y}}{Y} - s_L\frac{\dot{L}}{L} - (1 - s_L)\frac{\dot{K}}{K} \tag{2.6}$$

TFP and MFP are the more general concepts. The Solow Residual is bound to the neoclassical assumptions. The term 'MFP' is used when the aggregates of capital and labour are split in more than one input factor. In research praxis, SR, TFP and MFP are often used synonymously (Tuomi (2004)).

A related concept is to use output per capita of the workforce, often called 'labour productivity', or ALP in short, which can be calculated by subtracting $\dot{Y}/Y$ on both sides of (2.5), using $s_K + s_L + s_X = 1$.[23]

$$\frac{\dot{Y}}{Y} - \frac{\dot{L}}{L} = s_K\left(\frac{\dot{K}}{K} - \frac{\dot{L}}{L}\right) + s_X\left(\frac{\dot{X}}{X} - \frac{\dot{L}}{L}\right) \tag{2.7}$$

Due to the restrictive assumptions of perfect competition and constant returns of scale, one might be sceptical of applying the neoclassical approach to the real world. Its shortcomings have to be taken into account when it is used for productivity decomposition and measurement.

[23] This same equation results from taking the ln of

$$\frac{Y}{L} = \frac{K^{s_K} L^{s_L} X^{x_X}}{L}$$

and differentiating it with respect to time t.

In the TFP growth term of (2.6) all factors and effects are summed out which could not be measured directly. This might include omitted variables, spillovers and network externalities, reverse causality, increasing returns of scale, imperfect competition, or changes in resource allocation. On the other hand, there will be misleading results if the input elasticity deviates from the measured (real) input share. Then, the neoclassical model is not true, so that (2.4) is violated. (For a more detailed discussion about the neoclassical foundations, see Sect. 2.4.3. Shortcomings regarding the measurement of the productivity effects of digital goods are discussed in Sect. 2.4.4.)

2.4.3 Productivity Accounting and Empirical Studies Regarding ICT

Three basic strategies are used to evaluate the effects of ICT investments on productivity growth, called the firm-, industry-, and macro-level approaches, which are presented in the following sections.

2.4.3.1 Firm-Level Studies

Firm-level studies employ a wide range of different methods, using case studies as well as econometric analysis of panel data, mainly based on labour productivity or production functions. Pilat (2004) gives an overview. He summarizes that the evidence 'suggests that the use of ICT does have positive impacts on firm performance and productivity...However, the evidence also suggests that these impacts occur primarily, or only, when ICT investment is accompanied by other changes and investments.' Skills and organizational changes are two of these factors. Innovation, competitive effects, experimentation, time lag (benefits of ICT only emerge over time), firm size and age are other factors analysed by the studies. An overview of the results is as follows.

- **Labour productivity:** Almost all studies with firm-level data show a significant relationship between ICT and labour productivity. Hempell et al. (2004), for example, find that ICT capital-deepening led to higher labour productivity in Germany and the Netherlands in the period 1994–1998. Furthermore, they found a spillover effect from ICT to innovation behaviour of firms. Using a sample of 13,000 plants and introducing computer networks in an estimated production function, Atrostic and Nguyen (2005) find that computer networks have a significant effect on plant-level labour productivity in US manufacturing. See Arvanitis (2004) for an overview of this topic.
- **Time lag:** Brynjolfsson and Hitt (2003) find that the influence of computer investments on multifactor productivity growth 'monotonically and substantially increases' from a one-year to a seven-year difference specification using a panel data set of 527 firms in the period 1987–1994. The measured long-term contributions of computerization are significantly above computer capital costs (up to a factor of 5 and more).

- **Organizational restructuring:** Brynjolfsson and Hitt (2003) interpret these lags as indicators that the growth effects of computerization go along complementary organization investments over time and underscore the GPT character of IT. Falk (2003) found that companies introducing ICT are 'much more likely to experience new organizational practices' in Germany in the period 1995–1997. In a case study, Hughes and Scott-Morton (2005) describe the use of ICT in a transportation company. They find that the impact of ICT becomes strategically significant when it is embedded in a relevant set of complementary assets and the company has a sound strategic direction.
- **Production stages:** In another case study, Bartel et al. (2005) show the consequence of IT use in CNC machines for productivity and business strategy. They used plant data, especially data of one US valve production plant with 331 different products. Their result: IT use in CNC machines raises productivity at all stages of production and changes business strategy towards extended customization of valves. They concluded that other studies only considering labour productivity underestimate the total returns to IT investments.
- **Skills:** In their case study, Bartel et al. (2005) also found that IT use in CNC machines increased the skill requirements of workers. Using cross-sectoral data, Autor et al. (2003) prove econometrically that computerization is associated with a shift of labour input from routine to non-routine cognitive tasks in the US during the period 1960–1998.

Further references to several firm-level studies can also be found in OECD (2004c). Comparing firm-level results with studies on a more aggregated level, Pilat (2004), p.56, argues that 'aggregation across firms and industries...may disguise the impacts of ICT in sectoral and aggregate analysis. This is also because the impacts of ICT depend on other factors and policy changes, which may differ across industries.'

2.4.3.2 Sectoral Decomposition of Productivity Figures: Industry-Level Studies

Industry-level studies measure labour productivity growth by carrying out two steps.

1. The industries are grouped according to their IT-relevance (e.g. ICT-producing, ICT-using, non-ICT; or manufacturing and service). ICT capital stock or capital services or shares of these related to output or hours worked can be used as a numerical indicator for this ranking.
2. Labour productivity growth rates of the industries and groups are measured with output (value added or gross output) and employment data. Productivity of the industries sum up to total labour productivity growth. A simple standard decompositon approach is based on:

$$\frac{Y}{L} = \sum_{i=1}^{n} \frac{Y_i}{L_i} \frac{L_i}{L} \tag{2.8}$$

Labour productivity for the whole economy equals the sum of the productivity contributions of each industry group i weighted with their labour share L_i/L. A detailed description of methods for productivity decomposition is offered by OECD (2001). Input shifts (e.g. labour reallocation) and trade in intermediates have to be considered, which can be done by comparing two time periods (Ark et al. (2002)).

The advantages of this method are that detailed comparisons between the sectors can be made, and that the significance of a group for total productivity growth can be tested econometrically. For example, Stiroh (2002) and Ark et al. (2002) extend their decomposition analysis with a simple econometric difference test using dummies for industry groups.[24] Stiroh (2002) tested 62 US industries for the significance of their IT share for output growth and find a positive, significant link between the IT share of capital services in 1995 and subsequent productivity gains in the period 1995–2002. This underpins the results of his decomposition analysis, but Botsch and Stiroh (2007) cannot confirm the link between growth in the period 2000–2004 and IT intensity in the year 2000. Ark et al. (2002) found that in Germany, for the period 1995–2000, the productivity growth in the ICT-using manufacturing sector was significant for total labour productivity growth, but not the ICT-using services, whereas in the US, both ICT-using manufactures and services have been significant. Table 2.12 offers an overview of the results of studies using the industry-level approach.

2.4.3.3 Macro-Level Studies and Sectoral Decomposition of TFP

The industry-level approach predominantly uses output and labour data, the macro-level approach mainly uses capital stock and price data. The steps to carry out the macro-level approach are:

1. Calculating labour productivity growth depending on input factors for the whole economy
2. Subtracting the contributions of capital-deepening and changes in labour quality gives
3. TFP growth rate as residual
4. The sources of TFP growth rates are analysed sector- or groupwise. Groups can be ICT-producing and ICT-using industries, for example. Estimations of TFP growth in each sector are done by sectoral production functions or by sectoral costs functions (duality approach)

---

[24] Stiroh (2002) models the Difference-in-Difference equation as:

$$dlnA^Y_{i,t} = \alpha + \beta D + \gamma C + \delta DC + \varepsilon_{i,t}$$

with $A^Y_{i,t}$ for the industry's i gross output labour productivity in year t and C = 1 if IT-intensive, and C = 0 if otherwise. Dummy D represents two different time periods. So $\beta + \delta$ represents the acceleration for IT-intensive industries after 1995.

Table 2.12 Estimated contributions to labour productivity growth

Selected results	Approach[a]	Labour productivity growth	(1) ICT Capital deepening	(2) TFP Growth	(a) ICT producing	(b) ICT using	(c) Non-ICT industries
Pilat and Wölfl (2004) Germany 1996–2002	1	2,11	//	//	0,35[b]	0,17	1,58
van Ark and Piatkowski (2004)[c]	2	1,1	0,4	0,3			
EU-15 1995–2001	1	1,34	//	//	0,58	0,46	0,29
Jorgenson (2005) Germany[d] 1995–2001	2	1,29	0,46	−0, 10	0,65		−0,75
Inklaar et al. (2005) EU-4 1995–2000	2	2,02	0,53	1,07	0,53	0,19	0,35
Eicher and Roehn (2007) Germany 2000–2003	2	1,57	0,29[e]	−0, 01	0,17	0,13	−0,31
Jorgenson (2001) US 1995–1999	2	2,11	1,24	0,75	0,50[f]		0,25
Stiroh (2002)[g] US 1995–2000	1	2,62	//	//	0,63	1,55	0,45
Oliner and Sichel (2002); US 1996–2001	2	2,43	1,19	0,99	0,76[h]		0,23
Pilat and Wölfl (2004) US 1996–2002	1	1,74	//	//	0,61[i]	1,29	−0,15
van Ark and Piatkowski (2004)[j]	2	2,20	0,7	1,1			
US 1995–2001	1	2,19	//	//	0,98	1,17	0,06
Jorgenson (2005) US 1995–2001	2	2,23	0,92	0,54	0,48	0,06[k]	
Inklaer et al. (2005) US 1995–2000	2	2,46	0,86	1,05	0,71	0,68	−0,34

[a] Approach 1: Measuring / Weighting Labour Productivity Growth in the sectorgroups directly; Approach 2: Estimating TFP and splitting up Total Factor Productivity to sectors.
[b] ICT-producing manufacturing: 0,17; ICT-producing services: 0,18
[c] van Ark & Piatkowski (2004) do not split up the TFP on ICT producing, using and Non-ICT-sectors, but they measured labour productivity in these sectors separately. So the rows 1,2 and a,b,c represent different calculation approaches with varying total labour productivity.
[d] In the study of Jorgenson (2005) also the results for Canada, UK, France, Italy and Japan are given. In the period 1995–2001 UK had the highest contribution to TFP from IT production, whereas US had the highest contribution to labour productivity growth from IT capital deepening.
[e] Total contribution of capital deepening to ALP: 1,14; Comparison: In period 1995–2000: total capital deepening 0,88 and ICT capital deepening 0,33
[f] Jorgenson decomposes TFP into contributions from 'Information technology' and 'non-information technology'.
[g] Labour productivity growth is weighted; decomposition using industry value-added-productivity; no prior decomposition in ICT capital deepening and TFP; contribution of hours reallocation: −0, 34
[h] semiconductors, computer hardware, software, communication equipment; Other sectors: 0,23
[i] ICT-producing manufacturing: 0,45; ICT-producing services: 0,16
[j] see footnote c.
[k] non IT-production

This approach is used, for example, by Oliner and Sichel (2000), Oliner and Sichel (2002), Jorgenson (2001), Jorgenson (2005), and Inklaar et al. (2005). Two studies using this approach are presented in more detail in the following. Formally, the steps are best described in Appendix 1 of Oliner and Sichel (2002). Different results using this method (see Table 2.12) are due to differences in time periods, classification of factor inputs, price measurements, countries or states compared, depreciation rates, calculation of capital stocks, etc., but all studies use the neoclassical model as an implicit assumption. Two examples of this approach are presented here.

The Oliner-Sichel Study

One of the most-discussed studies regarding ICT and productivity is Oliner and Sichel (2000). Their approach will be shortly characterized here. They updated the results later using the same methodology, extended by a 'steady state approach', with which they 'wish to estimate a plausible range for productivity growth in the future' (Oliner and Sichel (2002)), not discussed further here.

Oliner and Sichel (2000) basically use the neoclassical growth accounting approach represented by (2.5). They calculate the contribution to output growth from five inputs: computer hardware, computer software, communication equipment, other capital, and labour hours. They employ 'productive' stocks which weight investments of former periods in relation to up-to-date investment according to their efficiency.[25]

According to (2.5) the growth contributions are the products of the growth of input factor stock and the income share. The income share of an input factor i is approximated by:

$$s_i = \frac{(r + \delta_i - \pi_i) p_i K_i T_i}{pY} \tag{2.9}$$

with r for the real net rate of return, δ_i as the depreciation rate, p_i as the rate of price change of i, π_i measures any expected change in the value of the capital over and above the depreciation rate, $p_i K_i$ as the nominal capital stock, T_i as a composite tax parameter, and pY for total nominal output (or income). The real net rate of return r is common to all types of capital. The idea is, that the income share of input i is large enough to outweigh the depreciation, the taxes and the decline in prices of new computers and to earn the constant real net rate of capital, what is in line with the assumptions of the neoclassical approach. Otherwise, a reallocation would occur, so that the above equation is a result of a neoclassical 'equilibrium' approach. Oliner

[25] Oliner and Sichel (2000) assume that a PC with a 486 processor has one half of the efficiency of a Pentium. Taking the Pentium as numeraire, a 486 PC which has already been used three years (of a total depreciation period of four years for all PCs) would contribute 2/5 Pentium-units to the 'wealth-stock' but 1/2 to the 'productivity-stock'. The 'constant quality' price series for computers of the Bureau of Economic Analysis (BEA) are used to deflate nominal investment time series accordingly.

and Sichel (2000) assume that r is 4% p.a., which corresponds to the average return of equipment in the US.

They find that ICT contributes to total growth through use and production of ICT:

1. Use of ICT: The direct contribution to total growth of an input factor is the result of the income share of that input factor times the growth rate of that input. In the period 1996–1999, Information technology capital contributed 1.10% points to the total annual growth rate of 4.82%. Taking into account the growth rate of the labour force, the growth figures can be transformed into labour productivity figures, which is decomposed by the authors into capital-deepening effects (increasing capital stocks), growth in labour quality, and MFP (see also Fig. 2.10).
2. Contribution to MFP by production of ICT: The sectoral contribution to total MFP is at the core here. Oliner and Sichel (2000) suggest that the price index is an indicator of MFP: assuming that input prices of raw materials are stable, the MFP growth in semiconductor industries must be large enough to outweigh its rapid price decline. Taking into account output weights, an aggregate contribution to total MFP of ICT is calculated. Intersectoral relations (e.g. semiconductors as intermediates for other ICT goods) are cleared. The analysis shows that – although having a small output share – the ICT sector contributed two-fifths of the growth in non-farm business MFP during the second half of the 1990s, which is what mainly reflects the sharp price declines.

Taking these calculations together and correcting for growth in the labour force, Oliner and Sichel (2000) estimate that ICT accounted for about two-thirds of the surge in labour productivity growth between the first and second halves of the 1990s.

Gordon (2003) criticizes the results of Oliner and Sichel (2002, 2000): the productivity growth attributed to ICT use was overestimated. He argues that Oliner-Sichel assume an instantaneous payoff of IT investment in their study. Further, Gordon (2003) points to the contradiction that older establishments in the retailing sector show no productivity growth despite their extensive use of computer technology. Also, the gap between US and EU productivity figures in the retailing sector is a contradiction according to Gordon, since the same technology is used in both regions. Finally, Gordon argues that the cross-US evidence shows no clear correlation between growth and computer use.

While Oliner and Sichel (2000) focus on national US data only, other studies compare productivity figures between several countries. The study of Inklaar et al. (2005) is presented here as an example.

The Inklaar-O'Mahony-Timmer Study

An example of a study comparing productivity contributions between different countries is the work of Inklaar et al. (2005). The authors use data of 26 industries of France, Germany, the Netherlands, the UK (EU-4), and the US in the period

1979–2000. Conceptually, they follow a growth-accounting framework similar to (2.5) with the input factors as labour and six types of capital (three ICT-goods: computers, communication equipment, and software; three non-ICT-goods: transport equipment, other machinery and equipment, and non-residential structures). The prices for the ICT goods are deflated according to the US deflators. Labour quality growth is defined by the difference between the growth of labour input and the growth of total hours worked.

After building per capita numbers according to (2.7) Inklaar et al. (2005) calculate the contributions of input growth and TFP growth from each industry to aggregate labour productivity growth. Their results regarding ICT are[26]:

- **Contribution to labour productivity:** ICT capital-deepening contributes 0.53% points and TFP growth 1.07% points to the total of the economy's labour productivity growth of 2.02% in EU-4 in the period 1995–2002. For the US, the figures are 0.86 (ICT capital-deepening), 1.05 (TFP growth), and 2.46 (labour productivity growth) respectively.
- **ICT capital-deepening:** Within ICT-producing industries, the US leads in terms of ICT investment growth, while ICT investment in the telecommunication industry grows faster in the EU-4. Although overall growth rates of ICT investments are similiar, the ICT capital compensation makes up a much larger share of value added in the US than in EU-4, because the ICT stock is larger. Therefore, according to the basic neoclassical assumptions desribed before, ICT capital-deepening has a much larger contribution to aggregate labour productivity growth in the US than in EU-4.
- **TFP growth:** ICT-producing industries contributes 0.53 (0.71)% points, ICT-using industries 0.19 (0.68) and non-ICT industries 0.35 (−0.34)% points to TFP growth of 1.07 (1.05)% in EU-4 (the US) in the period 1995–2000. ICT-producing industries makes the largest contribution to TFP growth in EU-4 and the US, a result which is confirmed by Jorgenson (2005). The difference of 0.49% points is interpreted as a sign that EU-4 lagged behind the US in exploiting the benefits of ICT.
- **Change in TFP growth:** Comparing the two periods 1979–1995 and 1995–2000, the rise in TFP growth rates in ICT-using industries in the US is mostly related to three industries: wholesale trade, retail trade, and financial intermediation. In EU-4, these three industries has no effect on aggregate TFP growth. The TFP growth is mostly confined to industries that produce ICT goods and services. On a national level, some sectors show high TFP growth rates in EU-4, but this effect is cancelled by low or negative TFP growth in many other industries.

[26] Some more results are presented by Inklaar et al. (2005) in their study, e.g. the contribution to growth of non-ICT capital slowed down in EU-4 in the considered period. These results are not described further here due to the focus on ICT.

2.4.3.4 Interpretation and Limitations

Firm-level studies agree that ICT investments have a significant and positive impact on labour productivity growth (see Chap. 2.4.3). This picture becomes more differentiated when the industry-level and macro-level approaches are used. Table 2.12 gives an overview of the most recent results. In the upper half are the results for Germany and the EU, and in the second half are listed the results for the US.

Despite all the differences in methodology and criticism of basic assumptions, a pattern of ICT productivity effects can be discerned.

1. Investments in ICT had an enormous impact on productivity growth.
2. The relative importance of the ICT-using sector for productivity growth is higher in the US than in Europe. This can be interpreted as a sign that the diffusion of ICT in the US may occur faster (Ark and Piatkowski (2004); Inklaar et al. (2005)). Higher contributions to TFP growth by the ICT-using than by the ICT-producing sector is also considered as evidence for the GPT hypothesis.
3. Both methodologies, the industry-level and the macro-level approaches, lead to similar results.
4. Newer studies (>2000) show that:
   - The link between IT intensity and productivity growth becomes weaker after 2000 (Botsch and Stiroh (2007));
   - The contributions of the industry sector to average labour productivity have changed comparing the periods 1995–2000 and after 2000 (Eicher and Roehn (2007)) which explore 52 German industries;
   - Basu and Fernald (2007) point to the GPT character of ICT and argue that TFP growth depends on former investments in IT, whereas current investments in ICT lead to an underestimation of TFP. They see their theoretical model with complementary investments in organizational capital confirmed by the data of 1987–2004.

Further interpretation of the cross-country results is difficult, the interpretation of TFP figures being hypothetical to some extent. For example, Ark and Piatkowski (2004) also calculated TFP figures for the Central and East European countries (CEE 2.0% p.a. in 1995–2001), but argue that this is likely related to effects of 'privatization, restructuring, technology transfer, higher capacity utilization, improvement in managerial and business skills, an increase in human capital and more entrenched macroeconomic stability'. Nevertheless, they estimate that the relative contribution of ICT capital in the CEE countries 'would most likely still be higher' than in the EU-15 countries.

Other differences between the studies and the country-specific results might be due to problems and limitations which empirical studies are confronted with. The following list gives an overview in three categories.

1. Measurement problems:
   - Cyclical mismeasurement of inputs due to changes in capacity utilization, hours per worker and labour quality (Basu et al. (2003)).

- Non-rival and intangible assets such as databases with information on customers are not part of 'capital' or 'ICT capital stock' by definition. Because such intangible assets are the main source of increasing returns using ICT capital, investment is an inadequate proxy (Howell (2004)).
- GPT nature of ICT capital: Complementary investments in learning, reorganization and the like are necessary in the ICT-using sectors. These effects may lower contemporaneous TFP growth.

2. Country-specific differences:

- Differences in national accounts methodology. For example, the US is one of the few countries that have changed the measurement of banking output to reflect the convenience of automated teller machines (ATMs) (Pilat (2004)). The price measurement may be different, so the US uses a hedonic price measurement for computers and other ICT goods. But using the US price indices for all countries can also lead to problems (see the next point).
- Furthermore, Baudchon and Brossard (2003) argue that revisions of the US National Income and Product Accounts (NIPA) are responsible for differences in studies using older and more recent figures. In 1985, hedonic prices appeared; in 1995, chain-type indices were introduced; and since 1999, software expenditures are counted as ínvestments and not as intermediate consumption any more. Chain-type indicesuse prices of the relevant period and prices based on a fixed year. In an example, Gordon (2003) explores the dependence of productivity measurement from price measurement:

  In t1, 100 workers in the sector S1 produce an IT good with a value of 100. In the IT-using sector S2, 100 workers produce a final consumer good which has a value of 200. In t2, the IT good is capable of a 50% increase in consumer good output, and the value of consumer goods rises from 200 to 300. The number of workers in both sectors remains unchanged. At cost-based prices, the IT goods still have a value of 100 in t2, whereas with quality-adjusted prices the value is 200. Table 2.13 shows the resulting differences in productivity accounting. This example shows that hedonic prices favour the result that productivity gains are realized mainly in the IT-producing sector.
- Differences in specialization: Pilat and Wölfl (2004) emphasize differences in industrial specialization which come along with different price developments: only a few countries produce computers, where price falls have been very rapid, but not all countries calculate hedonic deflators as is done by the US,

Table 2.13 Price measurement and effects on productivity; Source: Baudchon/Brossard (2003), p.10

Sectors	Cost based methodology			Quality based methodology		
	S1 (%)	S2 (%)	Total (%)	S1 (%)	S2 (%)	Total (%)
Labour productivity gains	0	50	33	100	50	67
TFP gains	0	50	33	100	0	25

S1: producer sector; S2: user sector

Canada, and France. So using the US deflator, for example, could overstate the growth in the ICT-producing sectors of countries which are less specialized than the US in producing ICT goods. Inklaar et al. (2005) confirm that only some sectors are responsible for higher TFP growth rates through ICT capital-deepening.

- Degree of outsourcing: For example, a possible reason for differences in TFP growth rates of the US compared to Europe might be that 'the US is more advanced than the EU in transferring the low value added parts of its manufacturing industries to emerging markets' (EC (2005a)), a process which is not directly but indirectly bound to the use of ICT.
- Differences in regulations, especially product and labour markets, and differences in intensity of competition (Basu et al. (2003)).
- Cultural differences between countries: Erumban and Jong (2005) find that the 'power distance' and 'uncertainty avoidance' attitude are significant cultural factors that can explain some of the differences in ICT adoption rates between countries. A country has a high 'power distance' when the decision structures are highly centralized, authority and the use of formal rules being substantial.

3. Conceptual issues:

   - IT allows stronger customization of products so that companies remain competitive in their home markets. So total returns of IT investments are underestimated when looking only at productivity measure (Bartel et al. (2005)). The benefit of the consumers due to spillover effects is greater the more competitive a market is (Pilat (2004)).
   - Pervasiveness of IT: When IT is defined as relative measure (IT intensity), the distinction between IT-intensive and unintensive fails, when IT facilitates productivity growth in all sectors (GPT: this argument is an alternative explanation for the not-significant results of IT intensity for growth given by Botsch and Stiroh (2007)).
   - The neoclassical assumptions might not be true (see Sect. 2.4.4). Then production functions and/or mark-up prices have to be calculated explicitly (Brandt (2002)). Further, it should be noted that the implicit link between productivity and growth relies on the neoclassical model (see (2.6)). But cyclical effects and rigid markets may interfere with this link, which becomes relevant when considering short-term structural effects, e.g. consequence of digitalization for unskilled workers, especially.
   - If the actual input elasticity deviates from the measured (real) input share, the neoclassical model would not hold true, which leads to a violation of (2.4). Assuming

$$s_K = \frac{P_K K}{Y} < \frac{\partial Y}{\partial K}\frac{K}{Y} = \sigma_K$$

with s_K for the measured input share with $P_Y = 1$ normalized and σ_K for true elasticity. A smaller fraction of output is explained by the factor input of capital, so that the measured result for TFP growth rate is too large. Howell

(2004) argues that in the case of a constant wedge $w > 0$ between measured input share and true elasticity

$$\sigma_K - s_K = w$$

the measured TFP growth rate will exceed the true TFP growth rate depending on the growth of the capital stock:

$$TFPGrowth_{measured} = TFPGrowth_{true} + w\frac{\dot{K}}{K}$$

This conceptually gives a hint when capital-deepening leads to higher TFP growth. Interpreting K as ICT capital, an acceleration in building ICT stocks may also lead to higher TFP rates.

When evaluating the productivity impact of digital goods, some more difficulties arise, which are discussed in Sect. 2.4.4.

2.4.4 Productivity Measurement and Digital Goods

The empirical studies presented above consider ICT goods. On one hand, the concept of digital goods is narrower than the classification of ICT goods (see Sect. 2.2.4); on the other, exchanging digital goods over a network (the Internet) needs investment in ICT goods (servers, routers, etc.). So the productivity impact of investments in ICT goods could also give a hint for the impact of digitalization. Furthermore, Basu et al. (2003) interpret digitalization as a broader concept which also encompasses the organizational restructuring in companies. But the intangible complementary capital such as management know-how, organizational structures, etc., could not be observed directly in the data. Instead, indicators of 'IT readiness' or 'IT usage' can be observed.

Howell (2004) refers to the characterization of digital goods by Quah (2002) (see Chap. 2.2.3), and recommends change of TFP measurement, so that TFP is directly dependent on digital goods as an input factor. He analyses the consequences of Quah's characteristics of digital goods for productivity measurement (Quah (2002)).

- **Non-rivalry**, **infinite expansibility**, and **aspatiality**: Tracing digital goods should compensate for multiple use in different sectors and time periods, especially when development costs are accounted for in one sector only. The neoclassical assumption of profit maximization in every period does not hold. To cover multiple reuse across periods, Howell (2004) suggests a modified production function with

$$Y = F(K, L, X, D)$$

$$D = \int_{-\infty}^{t} D(s)e^{-rt}ds$$

so that the stock of digital goods D is treated as a cumulative stock. But problems remain, for example, how to find an appropriate price index to evaluate D or to consider the devaluation of digital goods by age in r.

- **Infinite expansibility**: Howell (2004) argues that the production functions must be explicitly observable. Indirect measurement based on prices (see (2.4)) is not adequate because infinite expansibility allows copies of the digital item at no cost, which means that no price is available, or the digital intermediates are bought for a price which is above the neoclassical marginal cost pricing.
- **Recombinant properties**: These also give a reason for direct measurement of the production function. 'Recombinant properties' means that the value of two combined digital goods is higher than the added value of the goods themselves. An example of a recombinant process is the accumulation of databases of personal sales histories. With every entry, the value of all information rises, which leads to a durable non-depreciating digital asset which is not captured by traditional productivity analysis. The production function of the combined good is non-convex, so the recombination is a source of increasing returns which does not go in line with neoclassical assumptions. See also the results of Hughes and Scott-Morton (2005) and Bartel et al. (2005) presented in Chap. 2.4.3.

The use of the neoclassical approach to measure productivity effects of digitalization may be misleading due to a lack of adequate prices, misspecification of the production function, non-observability of digital goods as input factors, and multiple use in different periods and locations. Or as Howell (2004) puts it: ICT goods process information, and information goods have economic characteristics which are not in accordance with the assumptions of the neoclassical model. The problem remains of finding appropriate indicators to measure inputs and outputs of the digital economy. Haltiwanger and Jarmin (2002) stress that the 'IT revolution is closely connected to the growth of sectors of the economy (e.g. services) that we have traditionally struggled to measure.' Inklaar et al. (2006) confirm that different productivities in market services and not goods-producing industries are the main culprit for productivity differentials between Western countries. Barbet and Coutinet (2001) emphasize that a 'more content-oriented approach of ICT' was essential to cover the digital economy statistically.

2.4.4.1 Software Accounting

An example is the measurement of the contribution of software, which can be interpreted as a digital good (see Chap. 2.2.4). Oliner and Sichel (2002) and others find that ICT investment influences TFP growth positively in the ICT sector itself. Software production might be one reason for this, because copies can be made at negligible costs. Output can be increased without a corresponding increase in inputs, which results in increasing returns and higher TFP figures.

Further, software is an example of the difficulties in assigning the investments to balance sheet positions so that productivity measurement leads to the correct results. Howell (2004) mentions three cases.

- **Single-use software:** Software is bought as an intermediate (e.g. an accountancy software). The investment is capitalized and appears in the balance sheet as ICT capital. Often, it substitutes labour (capital-deepening) or other capital. But the positive effects of scale of software production are bound to the software-producing company, not to the software-using company.
- **Machine-control software:** Software is bought at one time as a direct input in the production process (e.g. software to control a washing machine). The manufacturer uses a copy of the software in every machine. The copies of the software will not be shown in the balance sheet in the ICT (or other) capital, but the extra returns of the software reproduction are reflected in the (higher) output price of the washing machines.
- **Acquisition:** If the producer of the washing machine is bought by another company, the implicit value of the software will be part of the 'other' capital stock and not of the ICT capital stock.

These examples show that a digital good could also have an effect via non-ICT capital growth on TFP growth. Further, the method by which software is accounted for refers not only to an economic measurement problem, but also influences the possibilities of financing 'digital companies'. Welfens (2004a) points out that young and small companies in Europe are especially affected because of the bank lending-orientated financing system in continental Europe.

2.4.5 Summary and Open Questions

2.4.5.1 Common Sense

ICT goods have a positive impact on productivity. This is shown in firm-level studies and using data beginning in the second half of the 1990s and ending in the early years of this century. Industry-level and macro-level studies also show the impact of ICT goods on productivity, in the ICT-producing as well as the ICT-using sector, although national differences prevail.

Venturini (2006) rebuilds the stylized facts in a theoretical model with two sectors, an ICT-producing and an ICT-using sector. In the dynamic model, two spillover effects between the two sectors occur: investments in ICT goods, and consumption of ICT goods by customers, which improves their working skills (a learning-by-doing effect). His model theoretically describes the growth impact of ICT.

1. **Adoption effect:** Both sectors contribute to the acceleration in labour productivity due to an ICT capital-deepening.
2. **Production effect:** TFP rises because of relevant efficiency improvements of the ICT producers.
3. **Spillover effect:** The economy takes advantage of the indirect external effects of ICT usage by firms and households.

With modelling the learning-by-doing effect Venturini (2006) also points to an effect of ICT investments which can be best seen in firm-level studies. According to the 'new growth theory', which emphasizes the role of human capital for growth, the link between digitalization, human capital, and growth must not be underestimated. Unfortunately, in industry-level and macro-level studies, the quality of the workforce can only be measured indirectly. For example, Inklaar et al. (2005) use the difference between the growth of labour input and the growth of total hours worked as a proxy for labour quality, which might not be an appropriate indicator when considering necessary skills in a digital economy. On the other hand, structural effects also occur the other way around: in a survey of 4,150 German firms, Falk (2002) finds that a greater penetration of ICT is significantly negatively correlated with the proportion of medium-skilled and low-skilled workers. This may also change the international division of labour, an argument which is emphasized by Welfens (2007c).

2.4.5.2 Questions to be Solved

Some questions have been left open for future research, conceptually and empirically.

- **Temporary or Permanent Shock?** Not finally answered is the question whether the development of IT can induce temporary or permanent higher productivity growth rates. This question was raised specially in the late 1990s when the topic was discussed under the headline 'new economy'. Gordon (2000a) and Gordon (1999) advocate that the development of ICT was a temporary shock, whereas Oliner and Sichel (2002) expect higher growth rates in future. Jorgenson (2001) expects that after the Keynesian revolution, its undermining through stagflation in the 1970s by the New Classical Counter-Revolution led by Robert E. Lucas the 'unanticipated American growth revival of the 1990s has similar potential for altering economic perspectives'.

  But in a recent study, Jorgenson and Vu (2007) admit that input growth 'greatly predominates' productivity as a driver for world output. The contribution of IT investment has 'moderate substantially' in G7 countries since 2000, whereas the contribution of IT investment has continued to rise for developing Asia, Latin America, Eastern Europe, North Africa, the Middle East and Sub-Saharan Africa.

  In his argumentation, Gordon (2003) emphazises that the macro-environment has fostered ICT investment in the late 1990s, e.g. the stock market bubble and the low inflation rate. He concludes that productivity can continue to grow even if ICT investment continues to lag behind the unsustainably high growth rates of the late 1990s. He argues that the IT-related productivity surge is temporary, and that in a historical dimension, other inventions of the last two centuries (electricity, the internal combustion engine, etc.) have been more important for wealth (Gordon (2000b)).

  Klodt (2003) criticizes the argumentation of Gordon: The impact of ICT as a general-purpose technology is greater than the historical innovation of the second

Kondratieff cycle, because the diffusion is faster and broader. The arguments of Gordon are stuck in a too-narrow definition of what might be the consequences of using IT.

- **Role of Price Decline** The main argument of Gordon for his 'temporary hypothesis' is related to the price decline of computer power: it shows that its marginal utility slows down. As an example, Gordon mentions word processing, where the 'real revolution in word processing came at the beginning by substituting the typewriter'. He concludes that 'instead of waiting for the productivity boost to arrive, it is more plausible that the main productivity gains of computers have already been achieved' (Gordon (2000a, p. 65)).

  Tanaka (2004) argues the other way around, that the quality-based productivity gains are an improvement in real value produced which is reported as a reduction in the price of these products. 'Digital deflation is about quality, not quantity.' Therefore, digital deflation leads to higher demand due to better quality, which is a wealth-enhancing technical progress and not a sign of diminishing marginal returns.
- **Drivers for Diffusion** (sectoral and international) It has to be answered what the main drivers of international diffusion of ICT are, and how digitalization changes the classic economic sector. Empirical analysis can test which exogenous variables are significant for diffusion. Pilat and Devlin (2004) point out that costs and implementation barriers slow down the diffusion of ICT. Comparing 82 countries, Kiessling (2006) finds that institutional variables such as a democracy index are significant for the diffusion of the Internet. But institutions are not key drivers for international diffusion of ICT goods on their own. Therefore, the next part explores this question from a theoretical perspective.

2.5 Summary of Part I

Digitalization refers to the spread of digital goods and IT services over networks. The concept of dividing information into a string of binary figures for more reliable transportation over long distances was first used in the nineteenth century, with the use of the Morse code. At the end of the twentieth century, the digitalization of the economy has accelerated due to the development of the Internet, progress in IT technology, and a price decline of ICT goods. A good approximation of the consequences of digitalization on labour productivity offers the analysis of ICT investments. Firm-level, industry-level, and macro-level studies show significant impacts on labour productivity and total factor productivity growth, although the results differ in detail regarding considered countries, sectors, time periods and methodology used, e.g. the impact of ICT intensity on productivity growth seems to have become weaker after 2000 in developed countries. ICT-related studies are the best guess for the productivity effects of digitalization, since direct analysis of digital goods fails due to measurement problems and conceptual shortcomings regarding the neoclassical growth accounting framework.

Part II
Economic Theory

Chapter 3
Theoretical Foundations

In Part II 'Economic Theory', macroeconomic general equilibrium models are linked with aspects of digitalization. First, in Chap. 3, some theoretical foundations are presented with a focus on the microeconomic perspective. Then three macroeconomic models follow as shown in Table 3.1. The part 'Economic Theory' closes with a comparison of the presented models in Chap. 7.

Table 3.1 Structure of theory part

Topic	Microeconomic clues	General equilibrium models
Networks	Sect. 3.1.1	Chap. 4
IT services	Sect. 3.1.2	Chap. 5
Digital goods	Sect. 3.1.3	Chap. 6

3.1 Some Microeconomic Aspects

3.1.1 Networks

The standard approach of network economics postulates a U-shaped demand curve. First, the willingness to pay grows because every additional user of the network implies rising utility for all participants.[1] This 'network externality' is also called the 'bandwagon effect' and has severe implications for the pricing of network services.[2] Having reached a specific number of users, the willingness to pay declines due to technological effects (network congestion), or psychological effects (aversion to mass products).

With a flat supply curve two equilibria arise: P_1 as an unstable equilibrium, P_2 as a stable one. The standard approach, therefore, predicts a market equilibrium in the area of falling individual utility when an additional user joins the network.

[1] According to Bob Metcalfe, the founder of 3Com, a US company producing equipment for data networks, the value of a network goes up as the square of the number of users. In literature, this hypothesis is also known as Metcalfe's Law.

[2] For the bandwagon effects on telecommunication services, see, for example, Rohlfs (2005).

M. Vogelsang, *Digitalization in Open Economies*, Contributions to Economics,
DOI 10.1007/978-3-7908-2392-9_3,

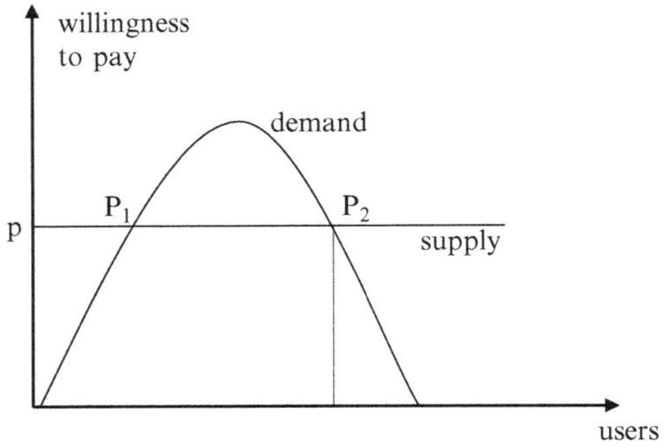

Fig. 3.1 Standard approach of network economics

A special form of network economics is described in the approach of two-sided markets. They are discussed in detail in Sect. 3.2.

3.1.2 IT Services

The development of a self-contained economic theory about services is just the beginning. An exception is the literature about spatial economics, which emphasize the role of services (Jansson (2006)). Of course, the standard supply-demand model may be applied, but the main differences between ordinary goods and services remain:

- Services are immaterial.
- The service itself cannot be stored.
- Services are related to consumers more specifically than goods.

The first two arguments preclude the consideration of intertemporal aspects. It is not possible to build a 'stock of services' in contrast to a physical or human capital stock, although it may be stated that the result of a service can be stored (see fnn. in Sect. 2.2.4 also). The third argument follows from the preceding ones: since production and consumption of the service (itself) cannot be separated temporarily, they are tailored more specifically to consumers than goods. From this it follows that services are produced with a higher labour-capital input ratio on average than goods, although exceptions occur, as Table 3.2 shows:

Since services are immaterial, the measurement of production, trade and consumption of services differs from the measurement of goods. In trade statistics, goods are measured when they cross the border. This concept is not applicable to services, where the balance of payment concept only is used. As a result, 'many trades in services are geographically domestic transactions made international solely by a difference in country of residence between the buyer and the seller of the service'

Table 3.2 Categories of service production plants; Source: Jansson (2006), p. 5

Main factor of service production	Recipient: persons	Recipient: goods
Labour	Day nursery	Cobbler's workshop
Capital	Solarium	Car park

(Lipsey (2006), p. 3). The different modes of trade in services are shown in Fig. 8.1. Further, services can be embodied in goods exports and imports which lead to an indirect trade of services.

The share of services in output and employment rises when countries develop (Hoekman (2006)). Two reasons can be mentioned:

- Economies of scale are higher in goods than in service production.
- An increase in the variety of goods, demand for higher quality, and specialization lead to a rising exchange of services.

Between companies, the purchase of services is also called 'outsourcing'. Internationally, it can be differentiated between 'international outsourcing' and 'offshoring', as Fig. 2.9 shows.

In this thesis, IT services are in the foreground as they are related to the concept of digitalization. IT services have been defined as B2B services in Sect. 2.2.2. They have accompanying characteristics since they are associated with a good or a company. The latter aspect is stressed in the IT Administrations Rights approach, which is introduced in Sect. 3.4.

3.1.3 Digital Goods

The special characteristics of digital goods (see Sect. 2.2.3) come along with special microeconomic features. Analysing Internet economics, these are according to Varian (2004).

- Price discrimination: Customization and versioning can be observed in digital markets. The recombinant characteristic of digital goods directly supports these strategies. 'Customization' or 'personalization' is also called price discrimination of the first degree; 'versioning' with a fixed menu of prices as price discrimination of the second degree; and selling different versions to different groups at different prices price discrimination of the third degree. An example is Microsoft, which sells versions of its operating systems to different customer groups at different prices. Variations of versioning are used for different delays (e.g. publishing stock quotes), user interfaces, convenience, image resolutions, speed of operation, flexibility of use, annoyance (start-up delays of reminders), and support (Varian (2000), Shapiro and Varian (1998)). That leads, for example, to time-based pricing, disaggregation of previously packed content, or bundling of products. An analytical approach towards versioning in digital markets is given by Belleflamme (2005).

- Bundling: When adding an additional digital component with zero marginal costs, the marginal costs of bundling are negligible. Regibeau (2005) analyses bundles of software and hardware and found different business strategies, e.g. Symbian wants to gain revenues with smart-phones, not with the operating system, but vice versa with Microsoft. Apple sells its iPod as a software/hardware bundle, but NTT DoCoMo prefers revenues through IP traffic and does not require a licensing fee from content providers.
- Switching costs and lock-in: The switching costs depend on the type of digital product. They are crucial to determine a business strategy. For example, bundling printers and ink cartridges, which are incompatible with other printers, means to lock in the customers when buying the printer.
- Standards: As seen in the example with printers and ink cartridges, the question of standards is crucial for market structure, welfare, and profits. Furthermore, a company which could enforce an industry-wide accepted standard can earn windfall profits. Standards are also coordinated by national and international bodies and institutions (see also Sect. 9.2.2).

3.1.4 Market Structure

The economic characteristics of digital goods and networks (see also Sects. 2.2.3 and 2.2.1) may favour monopolization. The most important factors are shown in Fig. 3.2.

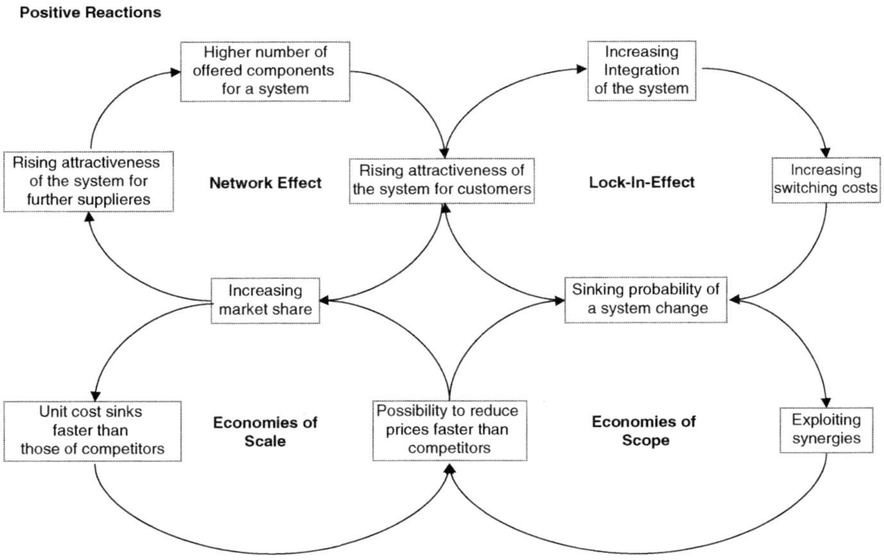

Fig. 3.2 Monopolization drivers; based on Müller et al. (2003), trans. and expanded by author

- Scale effect: Scale effects become relevant when marginal costs are close to zero, which is plausible with digital goods and ICT networks. Raising the quantities produced leads to lowering the average costs. This improves the competitive position of the expanding company.
- Network effect: Networks (but also selected digital goods or IT services) are more attractive the more the number of users are connected. See Sect. 3.1.1.
- Lock-in effect: Customers are bound to systems due to changing costs. Smith et al. (2000) refer to the similarity to loyalty programmes used by airlines and argue that familiarity with a retailer's site is a form of switching cost.
- Economies of scope: The recombinant characteristic of digital goods favours economies of scope, so they may go along with versioning (see Sect. 3.1.3). In the case of networks, economies of scope arise when TV, telephone and Internet access services are switched together by the access provider.

Although the role of versioning has not been finally proved (more niche markets may also attract more innovative competitors), the danger of monopolization is greater the more the effects enhance each other, e.g. scale economies may result in the survival of only one network among the conceivable alternatives (Liebowitz and Margolis (2002)).

In contrast to these factors, there also exist factors which favour competition. The most important are as follows:

- Love for variety, which means that a diversity of products is welfare-enhancing per se. Diversity can be related to B2B and B2C products, except homogeneous goods like telecommunication minutes. Standard microeconomic theory also relates love of variety to price discrimination: when preferences become more homogeneous, the welfare of consumers may rise despite price discrimination through versioning, since the pricing power of the supplier gets smaller.
- Altruistic and intrinsic behaviour to prevent monopolistic power, e.g. the open-source movement. See Sect. 2.2.4 for details.
- Schumpeterian view: Technological change that attracts innovative competitors. Monopolies are not (forever) sustainable, because 'creative destruction is the essential fact about capitalism' (Schumpeter (1942)).
- Institutional barriers, e.g. national tax laws or rules for invoicing and language barriers. These barriers might favour local monopolies, but prevent global ones.

Furthermore, Varian (2004) argues that monopolization tendencies with digital companies are not severe, because of the competition phase to acquire a monopoly implies low prices, IT fixed costs have reduced in recent years, competition with already-installed systems (competition with prior production), and pressure from complementors, who want to see lower system prices.

Also, it should be conceded that the previously-mentioned forces towards monopolization might not be all-dominant. Kling and Lamb (2000) point out that firms such as Amazon.com also could not have functioned on the Internet alone but rely on financial (credit cards) and distribution infrastructure. This limits market power. Agents in a network also act more heterogeneously than homogeneously.

Carley (2000) mentions three reasons. First, emergent behaviour, which means that the behaviour of groups, organizations, and markets cannot be predicted by looking at the average behaviour. Second, path dependence. What individuals learn is a function of what they currently know and who they know. Individuals with different backgrounds learn different things when faced with the same new information. Third, the inevitability of change: individuals learn continuously. That leads to changes in information, knowledge, and job networks. These arguments directly attack the proposition that networks drive monopolization.

3.2 Dynamics of Two-Sided Markets

3.2.1 Introduction

The theory of two-sided markets analyses platform markets with agents on both sides who interact through a platform. Examples are credit cards (shops and customers), operating systems such as Windows or Linux (software application producer and user), Internet search engines or portals, e.g. Internet auctioneer Ebay with sellers and buyers. This new strand of literature describes phenomena which are especially relevant for networks and digital goods.[3]

A short overview is given in the next section. The model presented in the following section describes a dynamic view of two-sided markets combined with the approach of the industrial economics of entry deterrence. As a result, the monopolist in the two-sided market initially invests an entry-impeding level of initial set-up and marketing efforts into its service. Then the two-sided network effects lead to fast growth of the platform, so that at the end, not only the technology and the customer base, but also the market position as a monopolist enhances the value of the platform provider.

[3] In the media, two-sided markets are recently highlighted by mergers and acquisitions of outstanding volumens. For example, the video portal Youtube was founded in February 2005. In November 2006, the company was acquired by Google Inc., a search engine provider, for Google shares worth USD 1.65 billion (SEC (2007)). The German student platform StudiVZ was founded in October 2005. In January 2007, it was bought by Holtzbrinck, a German publishing house. According to Spiegel, a German news agency, the price of the purchase was 85 million euro. The revenues of both companies, Youtube and StudiVZ, are low and very much below costs. Some more 'user-created-content' (UCC) platforms, which have recently been acquired by other companies, are listed in OECD (2007g).

Business newspapers have explained these deals as 'Web 2.0 hype'. 'Web 2.0' is a collective term for technological advances in browser-server-communications in the Internet and for sociological changes, since students (StudiVZ) and video-amateurs (Youtube) are ready today to publish private details on the Internet. For user numbers of three selected 'social network' Internet services, see ITU (2006a) for example. Technologically, the improvement is described by AJAX (Asynchronous JavaScript and XML), which is new, though not revolutionary from the technological point of view.

3.2.2 Two Sided Markets: Literature Overview

Two definitions are used to characterize two-sided markets:

- Cross-group network effects occur. The benefit enjoyed by a user on one side of the platform depends upon how well the platform does in attracting users on the other side (Armstrong (2006)).
- The volume of transactions between end-users depends on the structure and not only on the overall level of the fees charged by the platform (Rochet and Tirole (2005)).

The rationale behind the second definition is that the price structure determines the extent of cross-group network effects. In this sense, both definitions are equivalent, although Internet portals which do not charge users on either side do not come in the scope of the second definition. But in theoretical models, the second definition allows one to disregard the utility function and to start with the demand function directly. Economides and Katsamakas (2006) choose this way. They analyse application and operating system software and prove that 'network effects' in the sense of the two-sided market literature could also arise from the sales of complementary goods. Furthermore, the interactions do not necessarily refer to an exchange of a good or service between sellers and buyers. The platform can also be thought of as 'transforming values', e.g. students looking for information are asked for personal data which are used by the platform provider to target specific advertisements. The two-sided market literature is especially relevant for digital goods. Since they can be reproduced arbitrarily often without loss of quality at marginal costs close to zero, there is no 'expansion barrier' for production reasons for the supplier of a two-sided market.

Generally, platform providers can set four variables:

$$r_i, R_i \;\; i \in \{S, B\}$$

Transaction fees for sellers are r_S, and for buyers r_B; access fees for sellers are R_S, and for buyers R_B. Transaction fees are for using, and access fees for joining the platform.[4] Variable (or marginal) and fixed fees are synonymously used for transaction and fixed fees. Points out that the distinction between prices makes sense only if the choice of joining a platform is logically separated from the choice of making a certain interaction on that platform.

Two-sided market models use standard maximization techniques to determine the dependent variables. Usually, one of four different objective functions is chosen.

- Maximize the welfare of end-users: This strategy is used by Jullien (2006). Since end-users do not internalize the welfare impact of their use for the user on the

[4] Access fees could be split further into one-time registration fees and periodical fees. But the two-sided market literature is bound to static models yet, so the four above-mentioned variables have to be determined in one-period models.

other side of the platform, it is not surprising that welfare maximization requires a subsidization of fees.

- Maximize the profit of the platform provider.
- Maximize the social welfare, which is the welfare of the users plus the profit of the provider.
- Maximize the volume of trade.

Further, the models can be classified according to the number of platforms considered. In the case of more than one platform, the platform providers compete and the users decide on single-homing or multi-homing. In the case of multi-homing, the user is registered on or using several platforms simultanously. Assuming single-homing means inducing a higher degree of competition between the platforms. Further, 'single interaction' or 'multiple interaction externalities' could be taken into consideration. With single interaction, a single matching between two agents is realized, e.g. at employment agencies. Two-sided network effects arise when the quality or the probability of an interaction rises for members of one group when an additional member joins the group at the other side on the market. Multiple interaction externalities exist when every partner gets a benefit from each interaction, e.g. in Internet search engines. Peitz (2006) argues that the ownership of the platform also has to be taken into consideration. A private owner (e.g. a company) of a monopolistic platform provider or a high concentration of the ownership structure of several platforms could be socially beneficial since private owners are more inclined to internalize external effects on one side of the market.

Independent of the specific set-up of markets, the problem remains that an algebraic system with one objective function and four dependent variables is underdetermined, since r_i and R_i are subject to one demand respective utility functions of group i. To reduce the number of dependent variables, different techniques are used:

- Jullien (2006) analyses a situation without transaction fees first. He can solve the system with two objective functions (for buyers and sellers). This two-stage procedure implicitly leads to the result that registration fees should be used to determine the number of users, and transaction fees to determine the volume of trade.
- Rochet and Tirole (2005) introduce a new variable 'per-transaction fee' which combines transactions and access fees and acts like an average cost variable. Then the Lerner Condition is used, which relates the 'per-transaction fee' variable to the inverse of the elasticity of demand. Finally, the price structure (the relation between R_i and r_i) is determined by the requirement of identical marginal costs in each direction. But Rochet and Tirole (2005) admit that 'some redundancy in the pricing policy' remains when using this approach.
- Armstrong (2006) analyses a model with two platforms, single-homing agents and a 'two-part tariff'. With 'two-part tariff', the agents pay a fixed charge and marginal price for each new agent on the other side of the platform. So assumptions about the number of transactions are avoided. With eight dependent variables (two platforms, two sides, two prices), Armstrong (2006) shows that

multiple equilibria result. Which equilibrium of prices is chosen then depends on additional assumptions about the coordination process between the platforms.

An alternative approach is used by Welfens (2007a), who analyses the market of TV advertisements graphically. Independent of the specific set-up, the following results of the two-sided market literature can be summarized:

- Profit-maximizing might imply that the platform provider subsidize one side of the market. This happens when the demand elasticity of this side of the market is high or the external benefit enjoyed by the other group is large, e.g. directories are given to telephone subscribers free.
- With one or more platforms, positive cross-group network effects reduce profits, since platforms have an additional reason to compete hard for market share.

Finally, it should be added that Armstrong (2006) analyses competitive bottlenecks with the instruments of two-sided markets. Integrating a single- and multi-homing side in a model, he can explain why mobile companies compete strongly for new subcribers, but ignore the interests of those who call other mobile networks.

3.2.2.1 Empirical Evidence and Anti-Trust Policy

Evans and Schmalensee (2005) give an overview of two-sided platform markets in several industries and point out some regularities between the industries:

- Monopolies and near-monopolies are uncommon for industries based on two-sided platform markets. An exception is Microsoft, which had a market share of 96% of revenue in client operating systems according to Evans and Schmalensee (2005).
- Multi-homing on at least one side is common. Evans and Schmalensee (2005) interpret this as a sign of horizontal product differentiation.
- Asymmetric pricing is relatively common, e.g. operating profits from many two-sided platform providers stem from one side.

On the other hand, Evans et al. (2005a) analyse computer-based industries and compare the markets of PCs, video games, PDAs, smart mobile phones, and digital content devices (Apple's iPod). They identify very different pricing strategies despite similar technology.

The following model describes the strategy of a provider when launching a new digital platform.

3.2.3 Model: Basic Set-Up of Two-Sided Markets Optimization

Company A_1 plans to enter a market with a new service-platform which is based on a proprietary technology. The service-platform is two-sided: an increasing number of buyers b implies a positive (external) network effect for the sellers s and vice versa.

When entering at T_0 the intermediary A_1 is monopolist, i.e. it is the only provider of such a platform due to its advanced technology. It is the goal of A_1 to maximize the net value of the platform (its company), viz., to set prices and to determine the appropriate initial marketing effort. Revenues are generated by registration fees for buyers R_b and sellers R_s and transaction fees for sales induced by buyers r_b and sellers r_s. Registration fees can be positive, zero, or negative (which would be subsidization for joining the platform). Transaction fees are always ≥ 0. Initially, marketing costs M are spent to attract an initial number of buyers b_0 and sellers s_0 and investment F is made to build up the platform.

I assume 'multiple interaction externalities', i.e. every buyer could, per se, be interested in interacting with every seller. The number of trades or interactions I in period t is given by:

$$I_t = H_b(r_b)b_t s_t + H_s(r_s)s_t b_t = (H_b(r_b) + H_s(r_s))s_t b_t \tag{3.1}$$

with the assumption that a seller transacts with every $1/H_s$th buyer, and every $1/H_b$th seller is a counterpart for a buyer, matching probability $0 < (H_s, H_b) < 1$. I assume $\partial H_b/\partial r_b < 0$ and $\partial H_s/\partial r_s < 0$, so that an increase in the transaction fees declines the propensity to trade on that platform. An appropriate example for this set-up would be the market for credit card payments. Please note that 'transaction' then does not refer to the basic trade between shop and user, but only to the use of the payment system.

The platform itself is a digital good in the sense of Quah (2002). It is assumed that there are no marginal costs of operating the platform: when transaction fees are charged by the platform provider, each use induces a positive profit. Each member of both user groups generates additional revenues. The platform provider charges r_s and r_b for every trade induced from the sellers and buyers respectively. The profits generated by transaction fees amount to:

$$\Pi_r = \int_{T_0}^{\infty} \frac{r_b H_b(r_b)b_t s_t + r_s H_s(r_s)s_t b_t}{e^{zt}} dt \tag{3.2}$$

with z as fixed discount or interest rate.

The registration fees are one-time fees paid by every new buyer and seller joining the platform.

$$\Pi_R = \int_{T_0}^{\infty} \frac{R_b \dot{b}_t + R_s \dot{s}_t}{e^{zt}} dt \tag{3.3}$$

If R_b and R_s are negative, new members are subsidized. It is assumed that all transaction and registration fees are constant over time.

At every point in time, additional buyers are attracted depending on the number of sellers on the platform and vice versa. New buyers and sellers are attracted to join the platform according to:

$$\dot{b} = \alpha(R_b)s + \gamma(s - b) \tag{3.4}$$

and

$$\dot{s} = \beta(R_s)b + \phi(b - s) \tag{3.5}$$

with $\alpha(R_b), \beta(R_s) > 0$, $\partial\alpha/\partial R_b < 0$, $\partial\beta/\partial R_s < 0$ and $\gamma, \phi \geq 0$. These two equations are pivotal for a two-sided market.

The rise in the number of buyers in a period ($\dot{b} = db/dt$) depends positively on the number of sellers s and negatively on the number of buyers already registered. In the case of credit cards, more users join the platform when many shops offer facility of payment with the card, and some users may leave the platform if $b > s$, e.g. there are many buyers (credit card as a mass product) but only few shops available for s. γ and ϕ describe preferences, $\alpha(R_b)$ and $\beta(R_s)$ depend on the registration fees for the group of buyers and sellers respectively. The assumption $\alpha(R_b), \beta(R_s) > 0$ is the mathematical equivalent of the first definition of two-sided markets given above. Furthermore, not only does the user realize a higher utility by an additional member of the opposite group, but the platform provider's profit from transaction fees also rises. The marginal profit generated with the last new member is a rising function.[5]

Pivotal for maximizing the net value is to make sure that A_1 remains a monopolist. Therefore, it incorporates the goal of deterring market entry of a potential competitor in its strategy. Having entered the market first, A_1 is a Stackelberg leader. This implies that competition takes place in quantities (Cournot Case), not in prices (Bertrand Case). With a second player in the market, the profit of A would decline. But in the Stackelberg leadership game, the leader can predict the behaviour of its potential competitor and is able to influence the reaction function of the potential competitor. In his two-stage game-model about the role of investment in entry-deterrence, Dixit (1980) has shown that investments in production capacity could impede the competitor from entering, and that this could be a profit-maximizing strategy for the leader. Here, Dixit's production capacity is reinterpreted as the number of members already registered on the platform. Predicting that a competitor could have developed a similar technology at T_1^m at the earliest, it is assumed that $s_{T_1} = s_m$ and $b_{T_1} = b_m$ with $T_1 < T_1^m$ are entry-impeding, with:

$$s_m = \sigma L \tag{3.6}$$

and:

$$b_m = \omega L \tag{3.7}$$

with $0 < \sigma < 1$ and $0 < \omega < 1$ defining all potential buyers and sellers located in the country as percentage of total population L. The parameters σ and ω may be determined through access to technology, age, etc.

This implication, that A_1 plans to acquire all potential users, may be called hubristic and contradicts the standard theory of monopoly pricing at first glance. But two arguments should rationalize the point. First, if all potential users are members

[5] This may be rationalized by the supposition that initial users of a new platform, e.g. pupils and students, are short of money, whereas members joining later are older and have a higher income.

of A_1's platform, acquisition of new customers by the competitor would be costly since it has to reward A_1's customers for their opportunity costs, which are the costs of leaving a platform with b_m buyers (s_m sellers respectively) and joining a platform with no members initially. The opportunity costs are higher the more the number of members who join the platform of A_1. Therefore, a huge number of members is the best instrument of A_1 to prevent the entry of a competitor. Second, the platform offered by provider A_1 is a digital good by assumption, i.e. marginal costs are zero. With zero marginal costs, the endless expansion of a monopoly is profit-maximizing.

Summarizing, the net value of the platform at T_0 is the sum of starting-up costs $-(M + F)$ and the discounted revenues over time. The maximization problem yields:

$$Max \int_{T_0}^{\infty} \frac{r_b H_b(r_b) b_t s_t + r_s H_s(r_s) s_t b_t + R_b \dot{b}_t + R_s \dot{s}_t}{e^{zt}} dt - (M + F) \quad (3.8)$$

subject to:

$$\dot{b}_t = \alpha(R_b) s_t + \gamma(s_t - b_t) \quad (3.9)$$

$$\dot{s}_t = \beta(R_s) b_t + \phi(b_t - s_t) \quad (3.10)$$

$$\{b_0, s_0\} = f_m(M) \quad (3.11)$$

$$s(t > T_1) = s_m, \quad b(t > T_1) = b_m, \quad T_1 \leq T_1^m \quad (3.12)$$

$$\dot{s}_t(t > T_1) = \dot{b}_t(t > T_1) = 0 \quad (3.13)$$

with f_m as a function which relates initial marketing effort M to the initial customer base $\{b_0, s_0\}$ with $\partial b_0/\partial M > 0$, $\partial^2 b_0/\partial M^2 < 0$, and $\partial s_0/\partial M > 0$, $\partial^2 s_0/\partial M^2 < 0$. This system has to be optimized by calculating the optimal values of M, R_b, R_s, r_b, r_s.

Since f_m is a convex function, there exists an optimum M^* which maximizes (3.8). Assuming for a moment that a $M^* > 0$ has already been found, the values of b_0, s_0 are given, and the above system is a standard problem of optimal control.

Integrating the differential equations (3.9) and (3.10) into the objective function (3.8) and building the Hamiltonian L with multipliers λ leads to:

$$L = \frac{b_t s_t (r_b H_b(.) + r_s H_s(.)) + s_t (R_b[\alpha(.) + \gamma] - R_s \phi) + b_t (R_s[\beta(.) + \phi] - R_b \gamma)}{e^{zt}} dt$$
$$-\lambda_1(\alpha(R_b) s_t + \gamma(s_t - b_t)) - \lambda_2(\beta(R_s) b_t + \phi(b_t - s_t)) \quad (3.14)$$

The necessary conditions to maximize (3.8) are:

$$\frac{\partial L}{\partial r_b} = 0; \quad \frac{\partial L}{\partial r_s} = 0 \quad (3.15)$$

$$\frac{\partial L}{\partial R_b} = 0; \quad \frac{\partial L}{\partial R_s} = 0 \tag{3.16}$$

$$\frac{\partial b_t}{\partial t} = -\frac{\partial L}{\partial \lambda_1}; \quad \frac{\partial b_s}{\partial t} = -\frac{\partial L}{\partial \lambda_2} \tag{3.17}$$

$$\frac{\partial \lambda_1}{\partial t} = \frac{\partial L}{\partial R_b}; \quad \frac{\partial \lambda_2}{\partial t} = \frac{\partial L}{\partial R_s} \tag{3.18}$$

taking into account the transversality condition:

$$[L]_{t=T_1} \geq 0, \;\; T_1 \leq T_1^m, \;\; (T_1 - T_1^m)[L]_{t=T_1} = 0. \tag{3.19}$$

Since the condition (3.15) leads to:

$$\frac{\partial L}{\partial r_b} = \frac{H_b(r_b)b_t s_t + r_b \frac{dH_b}{dr_b} b_t s_t}{e^{zt}} = 0$$

$$\frac{\partial L}{\partial r_s} = \frac{H_s(r_s)b_t s_t + r_s \frac{dH_s}{ds_b} b_t s_t}{e^{zt}} = 0$$

and the term $b_t s_t (r_b H_b(r_b) + r_s H_s(r_s))$ vanishes in conditions (3.16)–(3.18), this allows further simplification of the problem. The resulting system (3.16)–(3.18) is a system in four differential equations with four variables $R_b, R_s, \lambda_1, \lambda_2$. But because the functions $\alpha(R_b)$ and $\beta(R_s)$ are not known explicitly and the choice of M is endogenous, the system cannot be solved unambiguously. Now, to have a closer look at the dynamics of the two-sided market, the evolution of the platform is split into three phases in the next section.

3.2.4 Model: Development of a Business Strategy

3.2.4.1 Splitting the Phases

As before, it is assumed that A_1 reckons that an imitator might build up a similar platform till T_1^m. This induces A_1 to split up its strategy into the three phases shown in Fig. 3.3.

In the beginning, the platform provider spends set-up and marketing costs. After market entry, additional customers 'automatically' join the platform, meaning through network effects only. Then, if market dominance was reached at T_1, economy of scale effects would be at work so that the threat of market entry of a new competitor is minimized and the monopolistic rents can be exploited: the economies of scale of the platform technology lead to declining average costs and increasing persistence of users to stay on the platform. Therefore, potential competitors will be deterred from entering after T_1. Overall, the strategy can be described as 'priority to market dominance'.

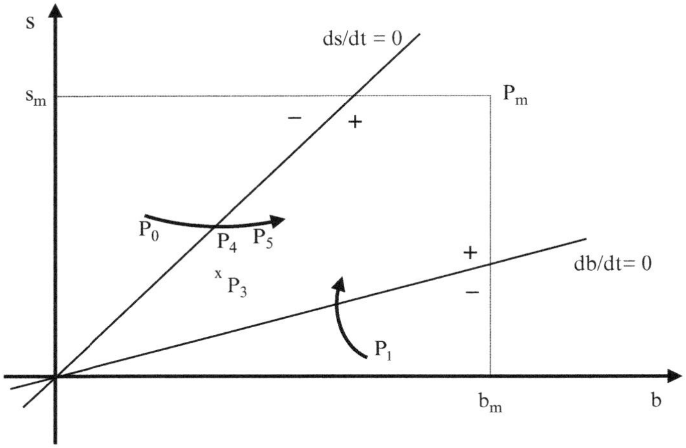

Fig. 3.3 Dynamics in phase 2

Before market entry (before T_0), the intermediary A_1 carries out these steps to develop the details of that strategy.

1. Determining transaction fees and potential number of customers b_m and s_m for phase 3 beginning T_1.
2. Calculation of the growth path of customers b and s between T_0 and T_1.
3. Calculation of the necessary number of customers b_0 and s_0 at t = T_0 to reach the growth path.

s_m and b_m are profit-maximizing because of (3.2) and because they underpin the monopoly in the long run.

Now, due to the split in strategy phases, profit maximization is not a dynamic problem any more, but a structural one to impose appropriate transaction and registration fees for sellers and buyers. De facto it works similar to a pre-commitment of the agents.

Phase 3: Exploiting Monopolist Rents

Monopolistic rents are exploited in phase 3, beginning at T_1.

However, in phase 3, the relevant demand function does not describe the dependence between a price and the number of users, but the dependence between a price (the transaction fee) and the number of transactions from users which are already registered at T_1.

Since fees and number of users then do not change during phase 3, the maximization problem of (3.2) could be simplified to:

$$\underbrace{max}_{r_b, r_s} \pi_r = r_b H_b(r_b) b_m s_m + r_s H_s(r_s) s_m b_m \tag{3.20}$$

Table 3.3 Market dominance-first strategy of a platform provider

	Market phase	Prevailing instrument
Phase 1	**Market entry**	Marketing
T_0	'Launch of the platform' Acquiring initial number of customers b_0 and s_0.	
Phase 2	**Automatic growth path**	Subsidizing/omitting registration fees
T_0–T_1	Network dynamics at work: Number customers of the platform grows 'automatically' till market saturation is reached	
Phase 3	**Exploiting monopolist rents**	Transaction fees
$T_1 - \infty$		

with:

$$\Pi_r = \int_{T_1}^{\infty} \pi_r e^{-it} dt = \pi \int_{T_1}^{\infty} e^{-it} dt$$

The optimization is straightforward:

$$\frac{d\pi}{dr_b} = H_b(r_b) b_m s_m + r_b \frac{dH_b}{dr_b} b_m s_m = 0$$

$$\frac{d\pi}{dr_s} = H_s(r_s) b_m s_m + r_s \frac{dH_s}{ds_b} b_m s_m = 0$$

which leads to:

$$r_b = -\frac{H_b}{\frac{dH_b}{dr_b}} = -\frac{H_b}{H_b'} \tag{3.21}$$

$$r_s = -\frac{H_s}{\frac{dH_s}{ds_b}} = -\frac{H_s}{H_s'} \tag{3.22}$$

or:

$$\varepsilon_b = \frac{dH_b}{H_b} \frac{r_b}{dr_b} = \varepsilon_s = \frac{dH_s}{H_s} \frac{r_s}{dr_s} = -1$$

The two-sided character of the model can be seen by the equivalence of the elasticities ε_b and ε_s.

Having determined s_m, b_m, r_b and r_s, the platform provider can start to plan phase 2.

Phase 2: Automatic Growth Path

The most important goal for the platform provider is to attract s_m sellers and b_m buyers at T_1. Four variables have to be determined: the registration fees R_b, and R_s; and the optimal initial values for the 'automatic growth path' s_0, and b_0 which has to be reached at the end of phase 1. Transaction fees are set according to the profit-maximization of phase 3 done earlier.

Transforming the system of two differential equations (3.4) and (3.5) into matrices leads to:

$$\dot{b} + \gamma b - (\alpha + \gamma)s = 0$$

$$\dot{s} + \phi s - (\beta + \phi)b = 0$$

or:

$$J\mathbf{u} + M\mathbf{v} = \mathbf{g} \tag{3.23}$$

with:

$$J = \begin{bmatrix} 1 & 0 \\ 0 & 1 \end{bmatrix}, \ \mathbf{u} = \begin{pmatrix} \dot{b} \\ \dot{s} \end{pmatrix}, \ M = \begin{bmatrix} \gamma & -(\alpha + \gamma) \\ -(\beta + \phi) & \phi \end{bmatrix}, \ \mathbf{v} = \begin{pmatrix} b \\ s \end{pmatrix}, \ \mathbf{g} = \begin{pmatrix} 0 \\ 0 \end{pmatrix}$$

The characteristic equation to solve the system yields:

$$|yJ + M| = \begin{vmatrix} y + \gamma & -\alpha - \gamma \\ -\beta - \phi & y + \phi \end{vmatrix} = y^2 + (\gamma + \phi)y - [\alpha\beta + \beta\gamma + \alpha\phi] \tag{3.24}$$

Setting the characteristic equation = 0 and solving the equation for the roots y_1, y_2 leads to:

$$y_1 = \frac{-(\gamma + \phi)}{2} + \frac{\sqrt{(\gamma + \phi)^2 + 4\,[\alpha\beta + \beta\gamma + \alpha\phi]}}{2}$$

$$y_2 = \frac{-(\gamma + \phi)}{2} - \frac{\sqrt{(\gamma + \phi)^2 + 4\,[\alpha\beta + \beta\gamma + \alpha\phi]}}{2}$$

The expression inside $\sqrt{.}$ is positive, so the two roots are real numbers. Root y_2 is surely negative. Root y_1 is positive, since $\sqrt{(\gamma + \phi)^2} = \gamma + \phi$ and $4\,[.] > 0$. A solution of the characteristic equation with real roots $y_1 > 1$ and $y_2 < 0$ signals a saddle-point equilibrium. Testing the trivial solutions $s = 0$ and $b = 0$ shows that the system of (3.5) and (3.4) is mathematically in equilibrium with no sellers and no buyers. A particular integral, therefore, yields: $\begin{pmatrix} 0 \\ 0 \end{pmatrix}$, since every solution of (3.23) can be used as a particular integral.

For graphical analysis, (3.4) and (3.5) are transformed:

$$\dot{b} = 0: \quad s = \frac{\gamma}{\alpha + \gamma} b$$

$$\dot{s} = 0: \quad s = \frac{\beta + \phi}{\phi} b$$

with $(\beta(R_s) + \phi)/\phi > 1$ and $\gamma/(\alpha(R_b) + \gamma) < 1$. Figure 3.3 shows the $\dot{b} = 0$- and $\dot{s} = 0$-demarcation-curves. Mathematically, only a trajectory with a starting point in the upper left quadrant or the lower right quadrant (e.g. negative values of s(b) and positive values of b(s)) can lead to the saddle-point equilibrium.

In Fig. 3.3, ‘starting points’ P represent the number of sellers s_0 and b_0 acquired in phase 1 through marketing efforts. All starting points in the upper right quadrant (positive values of b and s, as shown in the figure) finally lead to rising values of s and b. For starting points between the demarcation lines (such as P_3), the numbers of buyers and sellers rise monotonically. Starting points left of the $\dot{s} = 0$-curve imply that the number of sellers will shrink first till an appropriate level of buyers is reached so that new sellers also join the platform. Similarly, starting points below the $\dot{b} = 0$-curve lead to a smaller number of buyers on the platform first.

Only one trajectory leads to the target point P_m in figure given the values of R_b and R_s. P_m has to be reached at $t = T_1$ at the latest, so an initial part of the trajectory can be excluded due to reasons of time. Assume that P_0, P_4 and P_5 would be possible starting points. Then, according to the cost-minimization strategy P_4 would be also preferred to P_5 since the necessary marketing effort is a rising function in b_0 and s_0. But there is no information whether P_0 or P_4 is preferred by the platform provider without additional assumptions about the costs of acquiring buyers and sellers initially.

The dynamics shown in Fig. 3.3 are due to specific levels of the registration fees. To explore the relationship between registration fees and initial marketing effort further, the path of the trajectory leading to P_m is calculated next. The trial solutions for the dynamic paths are:

$$s(t) = m_1 e^{y_1 t} + m_2 e^{y_2 t}$$
$$b(t) = n_1 e^{y_1 t} + n_2 e^{y_2 t}$$

The parameters m and n are determined by equation:

$$(yJ + M)\begin{bmatrix} m \\ n \end{bmatrix} = 0$$

which has to be true for both $y \in y_1, r_2$.

For y_1 follows:

$$\left(\frac{-(\gamma+\phi)}{2} + \frac{\sqrt{(\gamma+\phi)^2 + 4\left[\alpha\beta + \beta\gamma + \alpha\phi\right]}}{2} + \gamma\right) m_1 - (\alpha+\gamma) n_1 = 0$$

$$-(\beta+\phi) m_1 + \left(\frac{-(\gamma+\phi)}{2} + \frac{\sqrt{(\gamma+\phi)^2 + 4\left[\alpha\beta + \beta\gamma + \alpha\phi\right]}}{2} + \phi\right) n_1 = 0$$

For y_2 follow:

$$\left(\frac{-(\gamma+\phi)}{2} - \frac{\sqrt{(\gamma+\phi)^2 + 4\left[\alpha\beta + \beta\gamma + \alpha\phi\right]}}{2} + \gamma\right) m_2 - (\alpha+\gamma) n_2 = 0$$

$$-(\beta+\phi)m_2+\left(\frac{-(\gamma+\phi)}{2}-\frac{\sqrt{(\gamma+\phi)^2+4\left[\alpha\beta+\beta\gamma+\alpha\phi\right]}}{2}+\phi\right)n_2=0$$

From the four equations follow:

$$\begin{aligned}
m_1 &= \frac{2(\alpha+\gamma)}{\gamma-\phi+\sqrt{\cdot}}n_1\\
m_1 &= \frac{\phi-\gamma+\sqrt{\cdot}}{-2(\beta+\phi)}n_1\\
m_2 &= \frac{2(\alpha+\gamma)}{\gamma-\phi-\sqrt{\cdot}}n_2\\
n_2 &= \frac{2(\beta+\phi)}{\phi-\gamma-\sqrt{\cdot}}m_2
\end{aligned} \tag{3.25}$$

Introducing two variables A_1 and A_2 and setting:

$$n_1=A_1 \quad n_2=\frac{2(\beta+\phi)}{\phi-\gamma-\sqrt{\cdot}}A_2 \tag{3.26}$$

so that appropriate linear combinations can be built leading to the general solution:

$$\mathbf{v}=\begin{pmatrix} b\\ s\end{pmatrix}=\begin{bmatrix}\frac{2(\alpha(R_b)+\gamma)}{\gamma-\phi+\sqrt{\cdot}}A_1e^{y_1t}+A_2e^{y_2t}\\ A_1e^{y_1t}+\frac{2(\beta(R_s)+\phi)}{\phi-\gamma-\sqrt{\cdot}}A_2e^{y_2t}\end{bmatrix} \tag{3.27}$$

In the case that $\gamma+\sqrt{\cdot}>\phi$ (the denominator of the first row) is positive, whereas the denominator in the second row is negative. With the boundary conditions $b(T_1)=b_m>0$, $s(T_1)=s_m>0$ the parameters A_1, A_2 in (3.27) could be determined:

$$\begin{pmatrix} b_m\\ s_m\end{pmatrix}=\begin{bmatrix}\frac{2(\alpha(R_b)+\gamma)}{\gamma-\phi+\sqrt{\cdot}}A_1e^{y_1T_1}+A_2e^{y_2T_1}\\ A_1e^{y_1T_1}+\frac{2(\beta(R_s)+\phi)}{\phi-\gamma-\sqrt{\cdot}}A_2e^{y_2T_1}\end{bmatrix} \tag{3.28}$$

Since $y_1>0$ and $y_2<0$ follow from the solution of the characteristic equation (3.24), the first summand dominates the second one. With assumption $\gamma+\sqrt{\cdot}>\phi$ it results that $A_1>0$. The sign of A_2 depends on the specific values of s_m and b_m. Looking backward in time from point P_m in Fig. 3.3, the sign of A_2 determines whether the path towards P_m was convex or concave.

Phase 1: Market Entry and Optimization

Having calculated A_1 and A_2, (3.27) determines the 'necessary' number of initial customers s_0 and b_0 at t = 0 in dependence from the choice of R_b and R_s. The marketing effort to gain the initial customers is given by the inverse of (3.11):

$$M=f_m^{-1}\left\{b_0,s_0\right\}$$

Summarizing, the platform provider has to resolve a trade-off between initial marketing costs M and registration fee R_b and R_s. The higher the initial marketing effort, the higher is the initial number of users s_0 and b_0 and the higher could be the registration fees without endangering the aim of reaching s_m and b_m at T_1. This is at the core of the optimization process here. Further, the digital platform provider will verify that calculated marketing budgets are used efficiently, e.g. that P_0 in Fig. 3.3 will not cause marketing costs higher than P_4.

Now the supplier chooses one of these strategies to solve the trade-off.

1. Minimize costs in initial phases 1 and 2. Costs are the sum of marketing effort and subsidization (negative registration fees) for joining the platform. This strategy may be chosen when financial resources are limited.
2. Maximize profit in initial phases 1 and 2. Thus, registration fees must be > 0 to outweigh the initial marketing effort.
3. Reach s_m and b_m as early as possible. This strategy will be limited by financial resources.

3.2.5 Summary

Two-sided markets describe situations where the utility of the seller rises with the number of buyers and vice versa. They are especially relevant for digital goods. Here, the strategic set-up of a two-sided market supplier is analysed theoretically. A key result is that a trade-off exists between initial marketing costs and registration fees. The theory of two-sided markets is used in Sect. 6.3 again. Then the interdependencies between macroeconomic variables and the strategic decision to enter a market are in focus. Some consequences of two-sided markets for regulatory policy are mentioned in Sect. 9.3.4.

3.3 Theories on Internationalization

3.3.1 Key Drivers in Management Literature

3.3.1.1 Business Strategies and Internationalization

Regarding possible cross-border relationships at firm level, three possible situations may lead to an internationalization of business: the need (or the goal) to find new customers and to improve the market share; the mandate to accompany a (domestic) customer; or to outsource internal functions internationally (offshoring or international outsourcing). The relationships work in both directions: a domestic company also may win an outsourcing contract from abroad which means that it has acquired a (new) customer. So the first and second arrows in Fig. 3.4 describe an export, whereas the third arrow symbolizes an import of services.

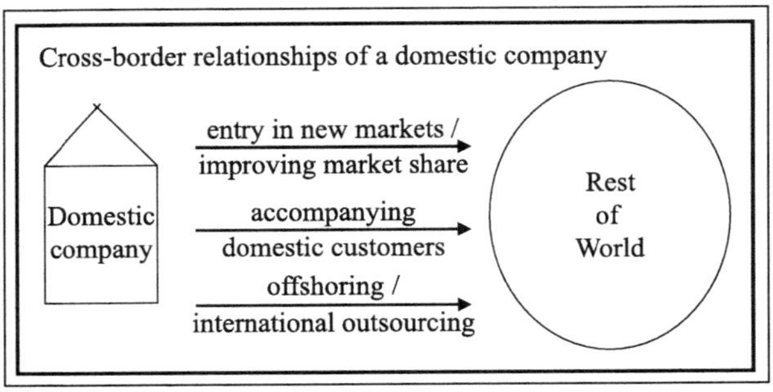

Fig. 3.4 Drivers of internationalization at firm level

Several approaches for internationalization strategies of companies are described in the management literature. They offer a glance at the strategic inner life of globalizing companies. Besides, entrepreneurship theories differ from most economic theories in that they to not assume coordinated behaviour between companies, but also accept that different strategies are sustainable and may lead to profitability. According to Meckl and Schramm (2005), the most important theories are as follows:

- **Monopolistic advantage theory** The multinational company has a unique advantage over other companies which is used for expansion.
- **Oligopolistic reaction theory** Oligopolistic market internationalization strategies are inherent, when the companies follow a forerunner.
- **Product life cycle theory** Companies export products to preserve sales opportunities if they are considered as mature (Vernon (1966)).
- **Step model of the internationalization** is dedicated to learning processes in organizations (Johanson and Vahlne (1977)). It stresses the risk aspects in foreign markets (e.g. moral hazard risks) which can be met by knowledge. Hence, a gradual entry into foreign markets takes place. Criticism is directed on three points: First, companies with huge resources can act more aggressively and may omit some steps, because the country-specific risk is relatively smaller. Second, experiences with foreign markets can be also collected in other ways. Third, with similar markets, an ad hoc entry may be rational if the market entry into one segment was successful.
- **Theory of internalization** is based on Coase (1937) and investigates different entry strategies like acquisition, joint venture or cooperation. In the internalization model, the entry mode choices describe the magnitude of control of the management. A franchise model has a low entry mode; an acquisition with entire integration of the management means having a high entry mode. Risk adjustment in the internalization model is done by the choice of the entry mode, in contrast to the step model of internationalization where the quantity and the speed at which resources are made available determines the risk exposure.
- **Network approach** The internationalization strategy of a company depends above all on its network, which represents the business potentials. Networks

in this context means contacts between companies and businessmen, but not networks in a physical sense as described in Sect. 2.2.1.
- **New venture theory** The new venture theory offers an explanation of the accelerating internationalization, while the importance of knowledge intensity is stressed. Quick and comprehensive market entry takes place to realize and to maintain knowledge advantages.

Looking at management processes, internationalization cannot be described by one comprehensive approach, also because spontaneous decisions by managers who react to business opportunities often play a big role in practice. Management decisions are always heterogeneous. Lacity and Willcocks (2001) mention that this might be particularly true for the IT sector.

- IT is not a homogeneous function. Some IT applications uniquely enable business operations and management processes. Others are less critical and favour the cross-functional integration of business processes.
- The pace of IT capabilities develops rapidly. Needs past a three-year horizon cannot be predicted properly.
- For efficient organization of companies, IT practice may be more important than economies of scale.
- Large switching costs between IT systems occur.

Gabrielsson and Gabrielsson (2003) argue that the drivers of internationalization are especially strong in the ICT sector. First, in the ICT sector, the requirements of worldwide markets are similar. Market segmentations take place on product or customer's groups, but not on regions. Second, only few trade barriers exist;, the telecom sector in particular is liberalized to a great extent. Third, strong competition in the ICT (and high tech) sectors requires that all advantages of global integration (scale effects) be exploited.

Analysing the history of eBay, an American online auction company, Mahnke and Venzin (2002) find the following characteristics of a digital company's internationalization strategy:

- Post-entry adaptations, i.e. additional modifications and amendments to the product can be carried out fast.
- Branding is more important in digital markets. Hence, rather high control modes are chosen, even if doubts because of foreignness and the danger of moral hazard in the foreign market seem to be low.
- The behaviour of customers in the foreign market can be identified and learnt more easily than in traditional markets. Thereby, a lower resource commitment is necessary.
- Investment costs are low (also with greenfield investment), so a long-term comparison may give the result that a greenfield investment or a merger and acquisitions transaction may have cost advantages against a joint venture with an unsuitable partner.
- The transportation costs are negligible with digital goods. Therefore, distance doesn't matter.

Due to these characteristics, strategies for companies with digital products seem suitable which deviate from the traditional internationalization strategies: market entry takes place faster and not step by step. Branding requirements lead to a higher entry mode, i.e. more extensive control, also in a country where the danger of being fleeced is low. Greenfield investments may occur rather than cooperations.

3.3.1.2 IT Outsourcing and IT Offshoring

IT outsourcing can be defined as 'obtaining information system services through external organizations which own some or all the necessary resources and where control and management of the resources and activities reside' (Jayatilaka (2002)). An overview of related definitions is given by Goles and Chin (2002). The EU definition of outsourcing and offshoring is shown in Fig. 2.9.

The management literature has evolved over time: First, the primary focus of outsourcing was on cost reduction. Then followed the idea of strategic outsourcing: managers should concentrate their resources on the company's core competencies only. Outsourcing nowadays is employed to redesign complete business processes (BPO, business process outsourcing) and increase value across the whole value-chain (Meyer and Weinert (2005)). This reflects that IT itself is regarded as a key enabler of the process-based organization (Lacity and Willcocks (2001)) and is managed as a strategic driver that generates long-term benefits and creates sustainable competitive advantages (Zhu et al. (2001)). The term 'transformation outsourcing' also describes organizational changes, but its concept is much more radical since it reshapes organizational boundaries totally, implying fundamental strategic and structural changes (Linder (2004)). Finally, comprehensive or total outsourcing characterizes the outsourcing of the whole IT system. Lacity and Willcocks (2001) recommend selective outsourcing because 'some activities can be outsourced while many others require management's attention, protection, and nurturing to ensure current and future business success'.

Some stylized facts on outsourcing are presented in Sect. 2.3.2. Regarding future developments, Carr (2005) predicts the most radical change in the IT sector. He compares IT with electricity and argues that electricity is not produced in a company today but in central power plants. He expects the same development to take place in the IT sector because the fragmentation was wasteful and capital utilization rates were low. Allweyer et al. (2004) argue similarly: the ubiquity of IT reduces its strategic significance. Bhagwati (1985) predicted in the 1980s that technological progress would lead to a splintering process which separates services from goods.

According to Lacity and Willcocks (2001), Quinn (2000), Allweyer et al. (2004), Erber and Sayed-Ahmed (2005), Rubalcaba-Bermejo (2004), Ang and Straub (2002), Meyer and Weinert (2005) and Dibbern and Heinzl (2002), the reasons for IT outsourcing from the perspective of the management literature are as follows:

- Reduction of costs, especially in small projects. A rule of thumb says that 300 users should use an ERP (enterprise resource planning) system so that an

in-house solution is economically advantageous (Allweyer et al. (2004)). The decision to outsource is based on comparative costs (production economics).
- Outsourcing is a strategic tool to complement missing internal capabilities.
- Exploitation of the full business potentials of new technologies to gain access to new knowledge and innovations and to improve existing IT systems.
- Outsourcing acts as a 'technical catalyst' to focus the organizational direction, also archieving partial or complete business transformation.
- The value chain is shortened, which diminishes risks (e.g. a new provider is responsible in case of offline times; service-level agreements are defined).
- It leads to more flexibility instead of being captured by long-term labour contracts.
- Saving of liquidity by lower investment costs.

On the other hand, the risks of IT offshoring and outsourcing include (apart from the authors mentioned above, see also Aubert et al. (2002), who give a short overview on IT outsourcing risks, and Kern and Willcocks (2002) who mention specific risks during contract building) the following:

- Unexpected high transaction costs for searching, creating, negotiating, monitoring, and enforcing a service contract between buyers and suppliers are considered. This might be true of international outsourcing especially: Erber and Sayed-Ahmed (2005) calculate in an example that the hidden costs may amount from 15% (best case) to 57% (worst case) of the original international outsourcing contract.
- Political risks, an insufficent enforceability of property rights, the risk of industrial espionage, language barriers, and different cultures of human resources management may cause trouble. Also, internal problems of balancing power.
- Outsourcing means loss of control and loss of know-how as well as dependency on an IT service company.
- Emergence of information asymmetries with the service company. This may be related to quality, additional and hidden costs, also to the vendor selection, the management of the organizational interfaces, layoffs and retentions, and managing the IT contract.
- High switching costs mean that the company may be locked in by an offshore relationship to some extent.
- Organizational issues matters. Short-term outsourcing decisions (maybe due to lack of capital) may lead to specific problems later. Also, IT cannot be isolated easily because many IT functions penetrate across business functions. Finally, opposition of the domestic workforce can lead to lower productivity despite international outsourcing.
- Business operations may be difficult: in a survey of Danish companies, Henten and Vad (2002) ascertain that 79% confirm that the delivery of their services requires that producers and buyers meet physically.

The different reasons for outsourcing mentioned here can be related to concepts used in management, and economic and social theories. A modified version of the classification of Lee et al. (2002) is shown in Table 3.4.

Table 3.4 Overview of outsourcing theories; based on LEE et al. (2002), expanded by author

Orientation	Theory	Focus	Resource	Main constructs
Strategic management view	Resource based	Internal slack resource	Physical capital, human capital, org. capital	1. Value 2. Rareness 3. Imperfect immutability 4. Nonsubstitutability
	Resource dependence	External resource (uncertainty)	Land, labour, capital, information, products (service)	1. Task dimensions (concentration, munificence, interconnectedness) 2. Resource dimensions (importance, discretion, alternatives, international availability) 3. Innovation
Economic view	Comparative costs	Cost-efficiency (also economics of scale)	Production cost	1. Production function a) Inputs: labour, capital, knowledge b) Productivity: Efficiency and organization
	Transaction costs	Cost-efficiency despite additional costs	Searching, nego-tiating, production, monitoring, delivery costs	1. Asset specificity 2. Uncertainty 3. Infrequency 4. Distance 5. Communication costs
	Agency cost	Principal-agent relationship (contracts)	Monitoring costs, bonding cost, residual loss cost	1. Uncertainty 2. Risk aversion 3. Programmability 4. Measurability 5. Length of relationship
	Technological change	Data/informaton exchange	Communication costs	1. Worldwide IP-network/ connectivity 2. Marginal costs of communication
Social view	Power political	Internal power-structure relationship	Power, politic	1. Power (Authority, resource acquisition, dependency and low substitutability, uncertainty absorption) 2. Politic (Selective use of decision criteria, selective use of information, use of outside experts, building coalitions, cooptation)
	Social exchange	Interaction processes	Trust, culture	1. Comparison level 2. Comparison level for alternatives

Table 3.5 Dynamics of offshoring; based on OECD (2004b), modified by author

	MNE subsidiaries	MNE subsidiaries providing outsourcing services to others	Local independents	Local joint ventures
Offshore activity	Supply insourcing services	Supply outsourcing services	Supply independent outsourcing services	Supply related party sourcing services
Initial stage	Driven by internal cost cutting at home	Compete to win outsourcing business	Compete for local subcontracting	Build on local knowledge to win market share of sourcing
Subsequent stage	Cut internal costs further to compete with rivals that are outsourcing	Compete with developing local independents to win outsourcing business	Establish subsidiary in country of origin of outsourcing	Restricted to local market or international spin-off

The resource-based theory implies that outsourcing is used to fill the gap between desired and actual capabilities of the company. Deficits in internal resources, knowledge or abilities are covered by outsourcing. The resource dependence theories in contrast focus on the long-term relationships of external companies. Comparative costs theories explain outsourcing by differences in labour, capital and knowledge costs or productivity. The transaction costs theory also takes into account the additional costs for searching, creating, negotiating, monitoring and enforcing a service contract. Dibbern and Heinzl (2002) deduce from transaction cost theory that the higher the degree of human asset specificity of an information systems function and the higher the degree of environmental and behavioural uncertainty is, the less likely is an outsourcing solution. This may also be true when the numbers of suppliers is small or the frequency of contracting is high. Agency-cost theories are related to transaction costs theories, but focus on the acting managers in the scope of the principal-agent approach. Theories which emphasize technological change focus on the costs and possibilities to build up a network. Finally, the social view is related to the resource dependence theories asking for the distribution of power or emphasizing the advantages of partnership (social exchange) respectively. The power theory also takes into account the internal power of the IT department, which might try to prevent outsourcing. Klein (2002) reviews a sample of the theoretical literature on outsourcing and finds that the transaction costs theory dominates.

A report of OECD (2004b) emphasizes the inherent dynamic of international outsourcing that 'competition has created a self-reinforcing dynamic'. It argues that an initial international outsourcing activity leads to further international outsourcing. Internal offshoring within a multinational company provides the confidence to begin international outsourcing. (See Table 3.5 for an overview of this dynamic concept).

3.3.2 Macroeconomic Theories on Internationalization

3.3.2.1 Traditional Approaches: Trade and FDI

Macroeconomic theory offers insights in basic driving forces of economies, and, thus, in the general conditions upon which internationalization decisions of companies are based. Traditional macroeconomic theories which describe internationalization patterns focus on trade or on foreign direct investment (FDI).

- **Trade** (purchasing at spot markets) In economic theory, the most important approaches are the model of Ricardo, explaning trade by comparative advantages due to different production functions and the model of Heckscher and Ohlin, which explains trade by different factor endowments of the countries considered. From the perspective of management literature, this implies that no or only a low integration of in-house processes with international partners is necessary. Newer approaches in trade theory modify the assumption of homogeneous goods and also allow for knowledge-spillovers. For an overview of modern trade theory, see Feenstra (2004).

  In a digital economy, the importance of knowledge rises, because its international diffusion occurs more easily. So the high-wage countries have to improve products 'climbing up the technological ladder' to earn higher returns in the sense of Schumpeter any longer (Welfens (2004b)).

  A special case of 'traditional trading' is the importing of services or intermediate goods, called international outsourcing or offshoring (see Sects. 2.3.2 and 3.3.1. Incorporating this case in theory has led to the theoretical models of fragmentation (see below). From the balance of payments view, it should be added that trade of services also includes the case that the customer moves to the service supplier (e.g. tourism, known as GATS mode 2) and the scenario that an employee of the service supplier is sent to the foreign country (part of GATS mode 4) (UN (2002)). See Fig. 8.1 for an overview about the GATS modes.
- **Foreign direct investment (FDI)** Foreign direct investment (FDI) changes the structure of property owners in the country invested in. It can be differentiated between:

  - greenfield investment, i.e. establishing a new plant in a foreign country; for an in-depth analysis of its determinants see, for example, Dunning (1992);
  - vertical integration, i.e. the acquisition of suppliers along the value chain (see Markusen (2002) for a comprehensive economic theory on this topic); and
  - horizontal integration, i.e. the merger of companies which are on the same position of the value chain; the emergence of multinational companies from a macroeconomic perspective is explained by Helpman and Krugman (1985).

  Regarding services, FDI corresponds to GATS mode 3: services are carried out by a local affiliate of the service supplier.

- **Migration** Migration describes cross-border labour mobility. Migration is covered by GATS mode 4. See Fig. 8.1 for an overview about the GATS modes.

Taking into account trade and FDI, both imply an international division of labour. Which pattern of specialization (if any) is chosen, depends on the goods, production and market characteristics, as well as on the relative factor endowments.[6]

3.3.2.2 Theory on Fragmentation

While traditional macroeconomic theories of trade focus on final goods only, a new strand of literature, the theory on fragmentation, also considers trade in intermediates. This approach is important for analysing service-orientated economies particularly.

The term 'fragmentation' refers to the decomposition of the production process in different locations. For instance, preproducts and intermediates are purchased to be assembled into a final product domestically. In the business management and economic literature, this and related phenomena are also treated under the terms outsourcing, delocalization, intra-product specialization, intermediate good trade, vertical specialization, splitting up of the value added chain, or production sharing (Feenstra (2004)). EC (2005a) uses the concept of 'relocation', which means 'the shifting of economic activities to overseas sites'. Besides, from a macroeconomic point of view, it is not important whether the intermediates are traded within a multinational company (MNC) or between independent companies, but trade along the production chain is analysed and modifies the predictions of traditional trade theory. The literature about fragmentation also addresses the following:

- Scale effects at pre-product or intermediate level: While scale effects are considered with the production of final products traditionally, the fragmentation theory allows scale effects at the pre-product or intermediates level to be taken into account. The higher the scale advantages are, the oligopolization or monopolization of this market segment can be expected. Therefore, the exploitation of scale effects on the one hand and the existence of several variants of a final product on the other hand is no contradiction within the scope of the fragmentation theory. To exploit economies of scale, it is also not necessary to organize complete production chains within a company any more. Intermediate and final goods producers both can profit from economies of scale within the value chain (Arndt and Kierzkowski (2001)).
- Factor income: The effects on labour and capital income by entering into worldwide trade become more unequivocal by fragmentation. In particular, the employees in a country whose firms shift the production of labour-intensive intermediates

[6] Zimny and Mallampally (2002), for example, find based on US data that the internationalization of services as a whole and of a majority of individual service industries takes place more through FDI than through trade.

to an offshore location will not necessarily have to accept real wage losses. Depending on the competitive intensity and the factor endowments of the countries, constellations with a high-wage and a low-wage country can also be sustainable in the long run (Jones and Kierzkowski (2001)). For an overview of empirical studies about the effects of trade on labour markets, see EC (2005a).

- Importance of communication networks: The IT sector plays an important role to coordinate the cross-border production processes. This aspect is pivotal in the model of Harris (2003), which is presented in Sect. 4. The volume of international trade in intermediate products grows when the (worldwide) network costs fall. Welfens (2001) emphasizes that investments in ICT could also cause outsourcing: if it is capital-augmenting, the ratio of wages to interest rates will increase, which will stimulate firms to relocate ICT production abroad.

Theoretical extensions of the fragmentation approach include the trade of company-related services such as marketing and design (Yomogida (2004)) or to consider the available service supplies as endogenous: Ho and Hoon (2003) postulate that outsourcing is promoted by the variety of services in the destination country.

3.4 IT Administration Rights

3.4.1 Introduction

Miozzo and Miles (2002) and OECD (2004b) emphasize the relationship between IT service companies and multinational enterprises: for many service industries, the initial stimulus of their internationalization was the rapid growth and global spread of multinationals in manufacturing sectors. Offering B2B services, the IT service companies follow the internationalization pace of their customers. In this section, the link between the organizational structures of companies, IT administration rights, and the basic choice between two IT system architectures is described. It is proved that the choice of the IT architectures coincides with the choice of the organizational structure of the company.

The model presented in this section is based on the property rights approach from Grossman and Hart (1986) and is similar to the idea of Brynjolfsson (1994). The model supports the result of Williamson (1985) that transaction costs of any economic activity are determined by the asset specificity associated with that activity (Ang and Straub (2002)). Grossman and Hart analyse the distribution of property rights between two companies with incomplete contracts. They also prove that a uniform distribution of the property rights, i.e. both companies remaining independent, can lead to economically efficient results. With their model, they offer an alternative to the transaction costs approach à la Coase.

3.4.2 *The Model of IT Administration Rights*

IT administration rights determine the access policy of the IT systems S_F and S_J from final good producer F and intermediate good producer J. The administration rights include, for instance:

- The control of access including the administration of users, e.g. the assignment of users to certain access rights;
- All system changes or expansions and changes in the data model;
- The possibility of accessing all relevant company data which are stored electronically, i.e. customer's lists, prices, etc.;
- All internal and external safety measures.

IT systems are a mirror-image of the enterprise production processes. Therefore, controlling the IT administration rights also allows the observation and control of the company's workflow. Thus, the IT administration rights are an essential subset of the property rights. The following model shows as a theoretical innovation that the choice of the IT system and the distribution of the property and IT administration rights take place simultaneously.

It is the common objective of intermediate good producer J and final good producer F to increase total productivity through better coordination of the operational processes between both companies. This concerns, for example, order processing, production, logistics, etc. The coordination is carried out by information technology. In the model, two possible IT architectures can be chosen to organize the data interchange between the companies:

1. **Separated systems** which use standardized interfaces for data interchange. This means that the 'point of transfer'[7] is standardized and documented openly as well as the exchange format itself being in common use.
2. **Integrated system**. When integrating the systems, proprietary solutions are often used.

The administration rights which are at the disposal of the companies in both these cases are termed A^F or A^J. The following cases are conceivable:

- $A^F = \{S_F, S_J\}$, $A^J = 0$ The final goods producer has the administration rights for both systems, or the integrated system respectively.
- $A^J = \{S_F, S_J\}$, $A^F = 0$ The intermediate goods producer has the administration rights for both systems or the integrated system respectively.
- $A^F = \{S_F\}, A^J = \{S_J\}$ Every company keeps the IT administration rights of its system. Data are interchanged through interfaces.

Intermediate goods producer J and final goods producers F invest for their IT systems amounts of K_F and K_J respectively. The cost function for producing the

[7] An example is the data interchange via an ODBC (Open Data Base Connectivity) interface.

intermediate product is termed C and depends on the investment volume of the intermediate goods producer in its IT system with:

$$C = C(K_J) \quad with \quad C' < 0 \quad and \quad C" > 0 \tag{3.29}$$

After having purchased the intermediate goods from the intermediate goods producer J, the final goods producer F sells it to its customers. Expenses for finishing, marketing and logistics are deducted. This generates a turnover R which depends on the investment volume of the final goods producer F in its IT system:

$$R = R(K_F) \quad with \quad R' > 0 \quad and \quad R" < 0 \tag{3.30}$$

If both producers could agree on a complete contract, the companies would distribute the IT administration rights ex ante as well as bindingly negotiate the investment volumes K_F and K_J and the split of profits. In this case, they would jointly maximize the profit function $R(K_F) - C(K_J) - K_F - K_J$.

However, de facto, this way is not possible. Grossman and Hart (1986) point out that in a contract between two companies, all eventualities can never be regulated. Therefore, the contracts are incomplete. The company which prevails in the case of a conflict has the residual rights at its disposal, i.e. the rights which could not be described completely by the negotiated contract.

Residual rights also exist in the area of administration rights. If, for instance, an interface solution is agreed upon between two companies, the actual investment volumes of a company in its entire system remain unobservable since the quality, consistency, safety and availability of data are influenced by the total inverstment of the company.

Therefore, it is assumed next that both companies negotiate an incomplete contract which only determines the distribution of the IT administration rights ex ante, because neither the investment volume in the IT systems nor the split of profits can be negotiated bindingly.

The business transactions between the companies may follow this time profile:

- Period 0: The type of the IT system architecture (seperated or integrated) and the distribution of the IT administration rights are determined.
- Period 1: Both companies choose simultaneously and independently of each other their investment volumes K_F and K_J.
- Period 2: The companies observe mutually the actual size of their IT investment volumes and decide whether they want to trade the intermediate goods between each other. At the end of the period, the companies negotiate about the split of the profit.

The split of profits follows a Nash-bargaining solution. For the negotiating position of the parties, the value of their outside options V are crucial. These describe the alternative revenues which can be expected if the intended deal does not work out.

By assumption hold $R(K_F) - C(K_J) > V^F(K_F, A^F) + V^J(K_J, A^J)$ with $V^F(K_F, A^F)$ for the outside option of the final goods producer whose value depends on the actual investment volume in the IT system and the received administration rights. $V^J(K_J, A^J)$ accordingly. The assumption means that the intermediate good is actually traded between both companies. For a stable solution, it is assumed that by increasing the investment volume by one unit, the marginal revenues will see a stronger rise in response. The marginal cost falls more strongly than the value of the outside options increases: $R'(K_F) > V_1^F(K_F, A^F)$ and $|C'(K_J)| > V_1^J(K_J, A^J)$ with V_1 for the first derivative of V with respect to the first operand K.

The value of the outside option depends on the distribution of the administration rights and IT architecture chosen between both companies for data interchanging, e.g. the interface solution or integration of the systems. It holds:

- With an integrated system:

  $$V^n\{K_n, A(S_F, S_J)\} > V^F\{K_F, A(S_F)\} = V^J\{K_J, A(S_J)\} = 0 \quad (3.31)$$

  with n = (F, J). As soon as an integrated system is built up, the individual investments of the companies in their IT systems are sunk in the entire system. Hence, if a company owns only the administration rights for its 'subsystem', the value of its outside option is zero.
- With standardized interfaces:

  $$V^n\{K_n, A(S_F, S_J)\} = V^F\{K_F, A(S_F)\} = V^J\{K_J, A(S_J)\} > 0 \quad (3.32)$$

  with n = (F, J). With the interface solution a separation of the partners is possible anytime without impairing the integrity of the systems, i.e. the IT systems remain functioning. The investments in the IT system do not become sunk costs. Hence, presuming symmetrical investment behaviour, the single value of the outside options of each company administering its own IT system meets the value of the outside option of one company, having received the administration rights for both IT systems.

The Nash-bargaining solution determines the actual profit from the relationship between both the companies. Taking into account the value of outside options for both companies[8] this leads to:

$$U_F = \frac{1}{2}\left\{R(K_F) - C(K_J) - V^F(K_F, A^F) - V^J(K_J, A^J)\right\} + V^F(K_F, A^F) \quad (3.33)$$

for the final goods producer.

$$U_J = \frac{1}{2}\left\{R(K_F) - C(K_J) - V^F(K_F, A^F) - V^J(K_J, A^J)\right\} + V^J(K_F, A^F) \quad (3.34)$$

[8] Please note that A^F can have the value of $A(S_F, S_J)$ or $A(S_F)$ depending on the architecture choosen. J respectively.

and for the intermediate goods producer, assuming that the 'access profit' in the brackets is shared equally. Assuming further that the IT investment costs are completely written off in the considered period, both companies choose their optimal investment volumes K_F^* and K_J^* in such a way that the profit functions after amortizations are maximized $(U_F - K_F) \Rightarrow$ max and $(U_J - K_J) \Rightarrow$ max respectively. For the final goods producer results:

$$K_F^* : \frac{1}{2}\left\{R'(K_F^*) + V_1^F(K_F^*, A^F)\right\} = 1 \tag{3.35}$$

and for the intermediate goods producer:

$$K_J^* : \frac{1}{2}\left\{-C'(K_J^*) + V_1^J(K_J^*, A^J)\right\} = 1 \tag{3.36}$$

3.4.3 Distribution of IT Administration Rights

Incomplete contracts and the split of profits à la Nash lead to distortions compared to the solution with complete contracts. With complete and binding contracts, the joint optimization calculation would have yielded $R'(K_F) = 1$ and $-C'(K_J) = 1$. However, a hold-up problem arises with the Nash-bargaining solution: the optimum investment volume of the intermediate goods producer decreases because now only 50% of the marginal costs are relevant for its profit maximization. And accordingly for the final goods producer. Hence, underinvestment occurs, except the borderline case when the value of the outside option exactly meets the value of the original investment.

Following the arguments of incomplete contracts, both companies try to hold the magnitude of the distortions as low as possible from the beginning. The negotiating positions, the value of the outside options at the end of period 2 are anticipated and the administration rights are already distributed accordingly in period 0.

In other words, the theory of incomplete contracts leads to a solution which is economically efficient with respect to the constellations possible.[9] The interdependent relationship between administration rights and the IT architecture chosen becomes apparent, leading to the following results:

1. When establishing an integrated IT system, only one of the partners should receive all administration rights. Proof: if both companies disposed of partial administration rights, the propensity to invest of one company could be increased by redistributing the administration rights, without affecting the other. For example, $V^J = 0$ and $V^F > 0$ with $A^F = A(S_F, S_J)$ could be a result.

[9] Theoretically, only this 'anticipated ex-post-2nd-best-solution' allows to incorporate incomplete contracts in general equilibrium models.

2. With a system which is based on standardized interfaces, both companies should hold the administration rights for their sub-system. Proof: if a company received all administration rights, then transferring administration rights to the other company would increase its propensity to invest without decreasing the investment readiness of the dispensing company.
3. Deviations from these solutions lead to disadvantageous economic side-effects. If, for instance, the final goods producer compelled an independent supplier to use not standardized, but proprietary interfaces, it improves its short-term negotiating power. But on the other hand, this strategy causes the intermediate goods producer to lower its investment volume, which negatively affects the operational profit of the final goods producer. The final goods producer has to solve a trade-off here: improving the negotiating position or promoting a cost-efficient production of the intermediate good.

The distribution of administration rights and the choice of the IT system architecture coincide. From this, it follows that a standardized interface solution is likely to occur in a spot market, where the relationship between the trading partners is short-term, often anonymous and homogeneous goods are traded. Following IT the administration rights approach, the companies keep their administration rights completely because the costs of changing the trading partner have to be small. Therefore, trading in spot markets relies on standardized interfaces (with or without data intermediaries). The value of the outside options are positive for both trading partners. But when the companies integrate or exclusively cooperate along the value chain, it is crucial from the property rights view to protect the knowledge needed specifically for the products. 'Exit costs' then are higher the more integrated the IT systems are. Hence, there is an incentive to favour proprietary IT solutions. Individual ERP and CRM systems[10] are more favoured by vertically-integrated companies due to the administration rights approach.

3.4.4 *Potential Caveats*

This approach, which relates organizational issues with information technology, is simplified by the choice between only two architectures. In reality, neither solution shall be observed in pure form; nevertheless, this theoretical approach gives important hints into economic trends in the real world. However, the approach can also be criticized from a model-theoretical view:

- Companies have more than one trading partner. Hence, several ways of interaction and data interchange exist in reality.
- The results may change in a multi-period context.

[10] ERP, enterprise resource planning; CRM: customer relationship management.

- The input factor 'trust' is not taken into consideration. Confidence between trading partners is a crucial precondition in order to raise total productivity. This may be true especially considering more than one period. Leamer and Storper (2001) point out that trust comes from relationships, particularly in the Internet age. Aspects such as trust and reliability may prevent changing the IT service provider even if standardized interfaces are implemented.
- Investments in IT implicitly have a double function in the presented model. They reduce the marginal cost or increase the marginal revenues. At the same time, they are necessary to build up the interfaces or to integrate the systems.
- The organizational consequences of choosing an IT architecture are not addressed. Kling and Lamb (2000) emphasize the importance of socio-technical support when an IT system is implemented. They describe one case where 'divisions fought cooperation with the new system, attacking its design, technical adequacy, and feasibility. This process dragged on for years, costing numerous hours of effort and meetings. Divisional accountants even attempted to sabotage the system ...'.

3.4.5 Alternative Formulation

An alternative formulation shows that the approach also works when only one party is requested to invest to rise total profits. Assuming a competitive market with free entry, company J has to pay a market price P to the independent service provider Q for IT services. Two types of IT administration rights are considered: $A(S_J)$ encompass the right of the producer J to have access to databases, e.g. to make data entries, etc. $A(S_Q)$ describes the right of the IT service provider to change, use, and resell the source code of the software, etc.

Company Q has two possibilities to spend gross earnings which are defined as revenues minus operating expenses: for dividends, or as investments H in human capital. In the long run, equilibrium in the free market condition assures that no 'dividend-orientated' company will survive in the market. Investments are necessary to stay in the market. But in the short run, contracts are incomplete and the split of 'gross-earnings' cannot be observed by company J.

Company J is interested in investments H because this leads to a 'better' IT system in terms of reliability, velocity and diminishes the amount spent for the internal IT department by D(H) with $D'(H) > 0$ and $D''(H) < 0$. A better IT system, provided by the IT service company, for example, reduces the work time needed for data entries at J.

Hence, the values of the outside options are:

- With a proprietary solution/integrated system:

$$V^J\left\{H_Q, A(S_J, S_Q)\right\} > V^J\left\{A(S_J)\right\} = V^Q\left\{H_Q, A(S_Q)\right\} = 0 \qquad (3.37)$$

As soon as a proprietary solution is built up, the investments of the IT service company are sunk. Hence, the value V_Q of its outside option is zero. The outside option $V^J\{A(S_J)\}$ of the producer also amounts to zero, because it relies on know-how of the IT service company.

- With standardized interfaces:

$$V^J\left\{H_Q, A(S_J, S_Q)\right\} = \max\left\{V^J\{A(S_J)\}, V^Q\left\{H_Q, A(S_Q)\right\} > 0\right\} \quad (3.38)$$

Both companies can have outside options with positive values. The investments are not sunk for the IT service provider. On the other hand, the source code, databases, etc. can be used later due to standardized interfaces by the producer J. If the administration rights were concentrated, the joint value could not be higher than the maximum of the particular administration rights.

The savings of J due to improvements of the IT system works like an extra profit which is split between J and Q by factor χ. Appropriate values of D, V^J and V^Q are assumed, so that a service contract between J and Q works out. The Nash-bargaining solution leads to:

$$U_Q = \chi\left\{D(H) - V^J(A^J) - V^Q(H_Q, A^Q)\right\} + V^Q(H_Q, A^Q) \quad (3.39)$$

Assuming further that the investment costs in human capital are completely written off in the considered period, the IT service provider chooses its optimum investment volume H_Q^* in such a way that the profit function is maximized after amortizations ($U_Q - H_Q$). For the IT service company results:

$$H_Q^* : \chi D'(H) + (1-\chi)V_1^Q(H_Q^*, A^Q) = 1 \quad (3.40)$$

The maximization of J is not shown here, because it has no direct influence on H due to the incompleteness of the contract. Equation (3.40) shows the underinvestment-problem (hold-up): factor χ diminishes the relevant marginal return $D'(H)$. This is outweighed by a preferable high value of $V_1^Q(H_Q^*, A^Q)$, which can be achieved by an appropriate distribution of the IT administration rights. In the case of a system with standardized interfaces, this means that the IT service provider remains independent. Compared to the joint solution $V_1^J\left\{H_Q, A(S_J, S_Q)\right\}$ the independency raises the value of the outside option.

On the other hand, with a proprietary system, the best way is to integrate the service provider Q into the IT department of J to create a positive value of the outside option $V^J\left\{H_Q, A(S_J, S_Q)\right\}$ of the merged company. The result is expressed in Lemma 2, which is presented below.

Table 3.6 IT architecture, property and administration rights

	(Multinational) Company, e.g. holding	Final and intermediate producer separated
Property Rights	Concentrated	Separated
Technology	Integrated IT system	Interface solution
Network	Intranet	Internet
Software	Proprietary solutions	Standard solutions
IT administration rights	Concentrated	Separated
Organization of IT services	Offered by a department of the holding	Offered by an independent service company

3.4.6 Summary and Outlook

However, this criticism does not change the main result: the choice of organizational structures and the type of the IT systems architecture coincide. Integrated companies also integrate their IT systems. So IT administration rights become a subset of the more general property rights. Table 3.6 summarizes the connections.

In the model of Chap. 5, which desribes the trade of IT services in a macroeconomic general equilibrium model, two conclusions are used particularily:

- **LEMMA 1**: The distribution of IT administration rights is a mirror of the distribution of the property rights.
- **LEMMA 2**: The organizational structure between an intermediate goods producer and an IT service company is on par with the organizational structure concerning the final and intermediate goods producer.

Chapter 4
Networks in a Fragmentation Model

The following macroeconomic model, which mainly follows the approach of Harris (2003), emphasizes the role of infrastructure and shows the relationships between communications costs, pattern of specialization, and the volume of trade.

4.1 Setup of the Model

The model describes the fragmentation of the production process and differentiates between intermediates producers who only produce for the local market, and intermediates producers who produce for the market of M countries. The countries are symmetric in their size, labour force, consumption preferences, etc. For supplying internationally, additional communications and coordination costs arise. The model outweighs the smaller economies of scale when producing only for the local market than for M markets with the additional costs of going global.

The intermediates produced are not perfect substitutes. The final product is assembled without costs from all intermediates available. The CES-production function of the final goods yields:

$$U = \left(\sum_{i=1}^{n} y_i^{\rho} \right)^{1/\rho} \tag{4.1}$$

From the perspective of the final goods producer, U may be regarded as a quality level of the final good. Alternatively, U can be interpreted as the utility level of the consumers which rises with the level of specialization (love of variety), which is represented by the number of intermediates used in the final product.

The goods-market is characterized by monopolistic competition.[1] The elasticity of demand ε is equal to the amount of the elasticity of substitution $\varepsilon = -1/(1-\rho)$.

[1] In theory, the assumption of monopolistic competition is an easy-to-use standard tool, although not without criticism. For example, Neary (2004) argues that the intermediate producers could merge and so enlarge their monopolistic power.

M. Vogelsang, *Digitalization in Open Economies*, Contributions to Economics,
DOI 10.1007/978-3-7908-2392-9_4,

It is characteristic for models with monopolistic competition that the price of the variants exceeds the marginal revenue by the mark-up factor:

$$\lambda = 1/\rho > 1 \tag{4.2}$$

In every country, d producers offer their intermediates internationally (e.g. on the domestic and international markets), and k producers only work for the local market. Every intermediates producer manufactures exactly one variant. The number of variants which are available internationally yields:

$$v = dM \tag{4.3}$$

The total number of different variants offered to the final goods producer of a country then is:

$$n = v + k = dM + k \tag{4.4}$$

According to the symmetry of (4.1) the source of the variants is without importance for the final goods producers. Defining $y_i = y$ as domestic demand for all variants in the goods markets are in equilibrium when demand equals supply. With x for the output of an intermediates producer who supplies only locally, and with x* for the output of an intermediates producer supplying internationally, the equilibrium conditions yield:

$$x = y \tag{4.5}$$

and

$$x* = My \tag{4.6}$$

In equilibrium, globally-acting intermediates producers offer M times the output of a locally-supplying intermediates producer. The asterisk * indicates international variables.

The cost function of the intermediates producers employs variable and fixed costs. The cost function for producing for the local market consists of marginal costs b and fixed costs Θ_{fix}:

$$\Theta_G(x) = bx + \Theta_{fix} \tag{4.7}$$

An intermediates producer who offers his products internationally has to bear internationalization costs c additionally. The total internationalization costs $C = cv$ are the sum of the costs for communication D and coordination αv^2. These are due for the coordination between the several producers of intermediate goods including consulting, translating, etc. The coordination costs rise more than proportionally with the number of partners involved in the coordination process. The total global communications costs D describe the costs D_m to build up the local part of the international communication (IT) network (e.g. the Internet) in each of the M countries and to connect the users D_v:

$$D = MD_m + dMD_v \quad (4.8)$$
$$C(v) = MD_m + vD_v + \alpha v^2 \quad (4.9)$$

The average internationalization cost per intermediates producer is first declining then rising in v:

$$c = \frac{C(v)}{v} = \frac{MD_m}{v} + D_v + \alpha v = dD_m + D_v + \alpha v \quad (4.10)$$

The cost function for an internationally-supplying intermediates producer then yields:

$$\Theta_G^*(x^*) = bx^* + \Theta_{fix} + c \quad (4.11)$$

Please note that the communications costs do not depend on the level of output x^*, which reflects the assumption of zero marginal costs of using IT networks. Formally, the derivate of c to x^* in (4.10) is 0.

Equilibrium

In monopolistic competition free entry leads to zero profits, since the fixed costs are financed by the mark-up between price and marginal revenues. The intermediates producers choose an optimal output, which equals marginal costs and revenues. The 'law-of-one-price' holds for all variants i so that equilibrium applies:

$$p_i = p = p^* \quad (4.12)$$

The price of the variants is given by:

- For intermediates producers supplying locally by the mark-up rule of monopolistic competition:

$$p = \lambda b \quad (4.13)$$

Due to the zero-profit condition the revenue px equals costs (4.7) in equilibrium:

$$px = bx + \Theta_{fix} \quad (4.14)$$

Using (4.13) gives the optimal output level x:

$$x = \frac{\Theta_{fix}}{(\lambda - 1)b} \quad (4.15)$$

- For intermediates producers supplying internationally from (4.11) and (4.13) results:

$$px^* = bx^* + \Theta_{fix} + c(v) \quad (4.16)$$

$$x^* = \frac{\Theta_{fix} + c(v)}{(\lambda - 1)b} \tag{4.17}$$

From (4.6) in equilibrium $x* = Mx$ must be true. The comparison of (4.17) and (4.15) shows that $x* > x$ is possible only if $c(v) > 0$. To cover internationalization costs the intermediates producers supplying internationally have to reach higher economies of scale in production. By rearranging (4.15), (4.17), and (4.6), it follows:

$$M\Theta_{fix} = \Theta_{fix} + c(v) = \Theta_{fix} + \frac{MD_m}{v} + D_v + \alpha v \tag{4.18}$$

which determines v endogenously.

Equation (4.18) is a technological arbitrage condition. The costs for locally- and internationally supplying intermediates producers equalize in equilibrium. Figure 4.1[2] shows the arbitrage condition with the number of internationally-supplying producers on the horizontal, and costs on the vertical axis. Due to the convex shape of c(v) two equilibria arise, assuming appropriate values of M, Θ_{fix} etc. v_1 and v_2 reflect a diverging number of internationally-supplying intermediates producers v which, also implies a changing number of locally-supplying intermediates producers k, which has to be determined next. For aspects of stability, e.g. which of the equilibria v_1 or v_2 in Fig. 4.1 are more likely to occur, see Sect. 4.2.

The total number of producers is restricted by the country-specific resources available. The supply of labour in each of the M markets is L^s, labour being internationally immobile. The labour market is in equilibrium when:

$$L^s = L^d \tag{4.19}$$

which should always be the case with wage $w = 1$ which is the numeraire in this model. The labour demand of a country yields:

$$L^d = kbx + dbx^* + k\Theta_f + dc(v) + d\Theta_f \tag{4.20}$$

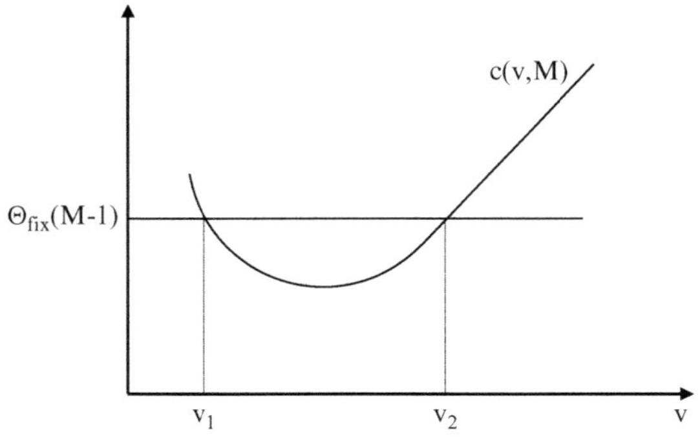

Fig. 4.1 Technological arbitrage condition, Source: Harris (2003), p. 62

[2] ©- by permission of Oxford University Press.

The current accounts of the countries are balanced. Without investments, the total demand for final products equals income with $wL = L$.

Since all international intermediates producers supply their products in all M markets, total revenues attainable with a single variant on a country's goods market yield:

$$py = L/n = L/(k + v) \tag{4.21}$$

Every intermediates producer gets $1/n$ of total demand of a country. Demand for final products is passed through to the intermediates producers directly.

From (4.15) and mark-up rule (4.13), it follows

$$px = \frac{\lambda}{\lambda - 1}\Theta_{fix} \tag{4.22}$$

The total number of variants n is determined by (4.21) and (4.22):

$$\frac{\lambda}{\lambda - 1}\Theta_{fix} = \frac{L}{n} = \frac{L}{k + v} \tag{4.23}$$

Having determined v already, (4.23) also determines k, so that the model can be solved fully.

4.2 Trade-Off Between International and Local Producers

Equation (4.23) postulates that the total number of intermediates producers n is given by a function of λ, Θ_{fix} and L, e.g. there is a unique solution for n. On the other hand, in Fig. 4.1, two possible equilibria for v are depicted, which go in line with two adequate equilibria for k, k_1 and k_2 with $k_1 > k_2$.

In both equilibria, the goods are produced with an equal value of internationalization costs:

$$\frac{MD_m}{v_1} + \alpha v_1 = \frac{MD_m}{v_2} + \alpha v_2, \quad v_1 < v_2$$

but their structure changes. The share of communications costs D in c(v) is higher in v_1 than in v_2. With v_1 producers, a market entry of one additional internationally-acting intermediates producer would lower the per-producer communications costs MD_m/v more than the rise of coordination costs αv. Profits become positive. More internationally-supplying intermediates producers are attracted: only locally-acting producers are squeezed out. From another perspective, former local intermediates producers start selling internationally. This development stops when v_2 is reached. A further market entry would now raise the coordination costs more than the fall of communications costs. Or, to express it in technical terms: v_2 in Fig. 4.1 is a stable, and v_1 an unstable equilibrium.

A borderline case arises when $v_2 > n$ with n from (4.23), so that $k = 0$. In this case, all only locally-supplying intermediates producers are squeezed out.

Or, as Harris (2003) puts it, the international economy becomes a global economy. With n internationally-supplying intermediates producers, a situation arises where profits of the producers are positive. New intermediates producers are attracted. Hence, supply of goods exceeds demand and the price p would fall below the profit maximizing level of monopolistic competition. This borderline case with n international intermediates producers leads to a disequilibrium in the terms of the model described.

4.3 Digitalization

Assuming, an international equilibrium is reached as described by point P_1 in Fig. 4.2 with $v_2 < n$. Then this model can be used to describe the macroeconomic consequences of digitalization. Digitalization changes communications behaviour (e.g. using email) and communications costs (e.g. telephone costs fall due to the introduction of digital transmission protocols (Voice over IP – see Sects. 2.3.4 and 2.3)). Therefore, digitalization induces falling communications costs D_m and D_v moving c(v) downwards to c'(v).

The reduction of internationalization costs attracts new competitors till P_2 is reached, and the technological arbitrage condition (4.18) is valid again. During this process:

- The number of internationally-supplying intermediates producers v rises;
- The volume of international trade rises although the output volumens x from (4.15) and x^* from (4.6) remain constant;
- The volume of international communication connections (number of telephone calls or volume of data transferred) rises;
- The number of companies with only local focus declines, since the total number of intermediates is limited by n according to (4.23) because of fixed labour supply L_s and constant mark-up λ;

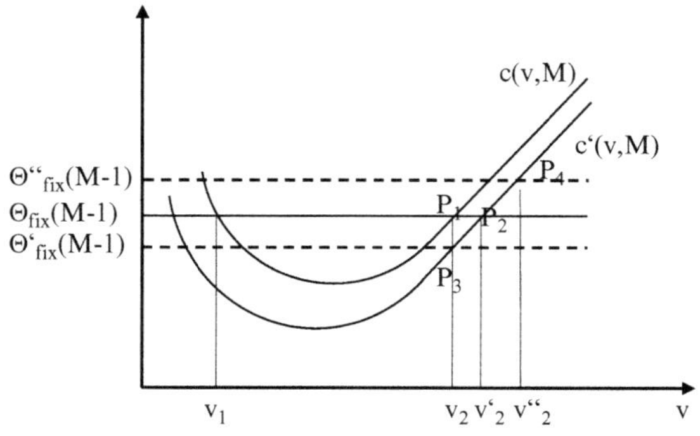

Fig. 4.2 Falling communication costs

- Hence, the relationship v/n rises, which could be interpreted as a measure of international fragmentation (or international outsourcing) in the production process.

Overall, despite the initial fall in the communications costs, the total global internationalization costs C(v) will have risen in P_2, since c(v) remains constant and v rises. Furthermore, C(v) and the volume of trade behave proportionally. The trading volume V_A of a country A results from multiplying the per-country revenues of the internationally-supplying intermediates producers with (M–1) markets:

$$V_A = d(M - 1)px$$

The worldwide trading volume V_{world} results by multiplying with M countries. With (4.22), (4.3), $\Theta_{fix}(M - 1) = c(\overline{v})$ from (4.18) and the definition of c(v) from (4.10) it follows:

$$V_{world} = d(M - 1)pxM = \overline{v}(M - 1)\frac{\lambda}{\lambda - 1}\Theta_{fix} = C(\overline{v})\frac{\lambda}{\lambda - 1} \quad (4.24)$$

with $\lambda = 1/\rho > 1$ results $\frac{\lambda}{\lambda-1} > 0$. $\overline{v}$ represents the equilibrium value of v determined by the technological arbitrage condition (4.18). Digitalization leads to lower communications costs and raises the global trading volume. The quality of the final product (the utility) is unaffected by the digitalization and expanding trading volume in this model: adding (4.1) for all intermediates with (4.21) and (4.13) yields:

$$U = n^{\lambda-1}\frac{L}{\lambda a} \quad (4.25)$$

Also, prices and revenues of the intermediates producers remain constant in the new equilibrium v_2' compared to v_2.

Leaving the equilibrium P_1 when digitalization has started, the producers may find it wise to deter new competitors from market entry. An appropriate approach would be to alter the cost structure of the intermediates production by innovations. Two strategies may be pursued:

1. **Limiting v – Home focus strategy:** The global companies try to reduce the fixed costs of production from Θ_{fix} to Θ'_{fix} in Fig. 4.2. A new equilibrium P_3 results where the output of the intermediates producers x and x^* declines and the number of intermediates producers who supply only on the local market rises. Trading volume falls, although the number of internationally-supplying producers v remains constant. Innovation in the production process leading to lowering of fixed costs Θ_{fix} allows production of more different variants of the intermediates since smaller economies of scale are necessary to cover all fixed costs. This is a macroeconomic pendant to the microeconomic argument that digitalization leads to versioning of products. This strategy is chosen when it is the aim of the already-existent internationally-supplying companies to limit an increase in v despite the decrease of x^*.

2. **Limiting n – Scale Strategy:** The global companies change production technology towards higher fixed costs Θ_{fix} and lower variable costs b. It results that quantities x, x^*, volume of trade V_{world} and the number of internationally-supplying companies v rise, whereas the price p and the number of local intermediates producers k fall. The new equilibrium is represented by point P_4 in Fig. 4.2. This strategy is chosen when it is the aim of the already-existent internationally-producing companies to lower the total number of competitors, which goes down by $(M\Delta k - \Delta v)$.

4.4 Interpretation

This model emphasizes the interdependencies between communications networks and international trade. It shows that digitalization favours the international fragmentation of the production process and fosters international trade. The number of international communications connections (e.g. the number of telephone calls or volume of data transferred) also increases. A high factor endowment of a country with labour raises the probability that locally- and internationally-supplying companies will coexist despite enduring falls in communications costs.

The main finding of the model is supported by the empirical literature. Freund and Weinhold (2003) find that an 1% increase in the number of Web hosts goes along with a 1% greater trade volume in 1998 and 1999. Welfens and Jungmittag (2002a) calculate that telecommunications has a positive impact on trade. The authors of both papers use the gravity model approach for their regressions, e.g. distance as one of the parameters to explain trading volume, but Freund and Weinhold (2003) could not prove that Internet had reduced the impact of distance. The empirical analysis listed in Appendix 12.3 also confirms the relationship between communications and exports: outgoing telephone minutes have a positive and significant effect on trade.

Chapter 5
IT Services in a General Equilibrium Macro-Model

5.1 Introduction

In this chapter, the relationship between trade in IT services and organizational forms in an international context is explored. The presented model is a General Equilibrium model with two countries, three sectors, and two organizational forms which is based on a model from Antras (2003). It allows linking of the factor endowments of a country with the architecture of the IT system and trade of IT services and uses the IT Administration Rights approach which is introduced in Sect. 3.4. The model emphasizes the accompanying character of IT services and shows that IT services are traded in both directions even when there are no differences in factor intensities or technology.

First, the general equilibrium model is described for an integrated world economy. Then, a two-country specification is considered, and, finally, trade in IT services is split into offshoring and international outsourcing.

The approach is in the tradition of newer macroeconomic General Equilibrium models which consider the choice of the organizational structure of a company as endogenous. In this strand of literature incomplete contracts, search costs, or incentive systems for managers are integrated to overcome the conclusions of Coase that the distribution of property rights is irrelevant from a macroeconomic point of view. According to the theory of Coase (1937), the distribution of property rights doesn't matter as long as they are distributed consistently. Contributions to this new class of General Equilibrium models have been developed by Grossman and Helpman (2005), Rauch and Trindade (2003), and Grossman and Helpman (2002), in which search costs play a central role, or the model of Antras (2003), which is presented in a modified version in the next sections. A comprehensive overview about the GE models with endogenous determination of the organizational structure is given by Spencer (2005).

M. Vogelsang, *Digitalization in Open Economies*, Contributions to Economics,
DOI 10.1007/978-3-7908-2392-9_5,

5.2 General Equilibrium in the Integrated Economy

5.2.1 Structure of the Model

The model consists of 2 countries, 2 factors, 2 industries with 3 sectors (IT services, intermediates, and final goods production), and 2 organizational forms.

The key assumptions of the model are as follows:

- Countries A and B only differ in their endowments of capital and labour.
- Trade of intermediates is possible; no trading costs occur. Factors are internationally immobile. Also, final goods are not tradable, so that final goods producers produce their varieties in all J countries. Cost of assembling the final product is zero.
- Each variety y(i) of the final good requires special and distinct intermediate inputs $x_y(i)$, $z_y(i)$ for variety i.
- There are two industries, Y and Z. Industry Y is more capital-intensive than industry Z.
- Each final goods producer decides whether to obtain the intermediate from a vertically-integrated supplier or from a stand-alone supplier.
- Investments in capital and labour are chosen simultaneously and not cooperatively by the final goods producer and its supplier. They cannot contract on them because no outside party can verify the amounts. The investments are useless outside the relationship. Only the allocation of the residual rights and a lump-sum transfer $T_k(i)$ is contractible ex ante.
- Incomplete contracts exist; bargaining leaves the final goods producer with a fraction ϕ, $1/2 < \phi < 1$ of the ex post gains from trade.
- IT services are used to organize the relationship between final goods and intermediate goods producers. They are necessary for each variant; IT services are produced with the capital-labour intensity of the world endowment. Total costs for IT services are proportional to the number of variants of the final good. The IT Administration Rights approach presented in Sect. 3.4 holds: the relationship between the IT service company and intermediate producer follows the organizational structure between intermediate and final goods producer.
- Free market entry: zero-profit condition also holds in General Equilibrium.
- Factor price equalization (FPE) holds, e.g. country-specific and world-relative factor endowments are 'not too different'. Equilibrium prices and aggregate allocations are those of an integrated economy.
- Consumers have identical preferences and love variety. They consider the varieties of each industry Y and Z as differentiated. They allocate a constant share μ, $0 < \mu < 1$ of their spending to sector Y, and $1 - \mu$ to sector Z. Decisions are not time-dependent, there is no discount of future revenues.

For a list of variables used, see Appendix 12.4.

5.2.2 *Basic Set-Up: Integrated Economy*

First, the integrated world economy is described. At the end of the section, the wage-interest rate ratio is determined which is valid in both countries due to factor price equalization, even if they are not integrated fully.

5.2.2.1 Demand

The preferences of the representative consumer are identical in both countries:

$$U = \left(\int_0^{n_y} y(i)^\alpha di\right)^{\mu/\alpha} \left(\int_0^{n_z} z(i)^\alpha di\right)^{\frac{1-\mu}{\alpha}} \tag{5.1}$$

with the elasticity of substitution between any two varieties in a given sector of:

$$\frac{1}{1-\alpha} > 1$$

and:

$$0 < \mu < 1$$

for a constant share μ which is spent for sector Y by the customers. The number of varieties n_y produced in industry Y and number of varieties n_z produced in industry Z are determined endogenously.

Due to the unit elasticity, the demand for two sectors Y and Z can be calculated separately. Following Antras (2003), the solution for an integrated economy is described next, so that country-specific indices are omitted.

With nominal total expenditure:

$$E = \int_0^{n_y} (p_y(i)y(i))di + \int_0^{n_z} (p_z(i)z(i))di \tag{5.2}$$

leads to a demand function for a variant i of industry Y products:

$$y(i) = \mu E \frac{p_Y(i)^{-\frac{1}{1-\alpha}}}{\int_0^{n_y} p_y(i)^{\frac{\alpha}{\alpha-1}} di} \tag{5.3}$$

with $p_Y(i)$ as the price for the variant and $P_Y = \int_0^{n_y} p_y(i)^{\frac{\alpha}{\alpha-1}}$ as a price index. The calculus is listed in Appendix 12.5 in detail.

Rearranging and splitting the price index in two sub-indices leads to:

$$y(i) = A_Y p_Y(i)^{-\frac{1}{1-\alpha}} \tag{5.4}$$

with

$$A_Y = \frac{\mu E}{\int_0^{n_{Y,I}} p_{Y,I}(j)^{-\frac{\alpha}{1-\alpha}} dj + \int_0^{n_{Y,S}} p_{Y,S}(j)^{-\frac{\alpha}{1-\alpha}} dj} \tag{5.5}$$

$p_{Y,V}$ for the price index for final products of integrated companies and $p_{Y,O}$ for price index for final products with separated intermediate producers.

In general equilibrium E also amounts:

$$E = rK^* + wL^* \tag{5.6}$$

with K^* and L^* as factor supply and r and w for capital (interest rate) and labour income (wage rate) respectively. The factors cannot be traded internationally, but labour and capital are homogeneous and flexible within a country, so that demand for them always equals supply in general equilibrium.

5.2.2.2 Production

Three sectors exist in the integrated economy: final good producers, intermediate goods producers, and IT services. The variable cost function for producing the intermediates $x_k(i)$ for variety i in industry k, $k \in \{Y, Z\}$ is determined by a Cobb-Douglas production function:

$$X_k(i) = \left(\frac{K_{x,k}(i)}{\beta_k}\right)^{\beta_k} \left(\frac{L_{x,k}(i)}{1-\beta_k}\right)^{1-\beta_k} \tag{5.7}$$

with $K_{x,k}(i)$ and $L_{x,k}(i)$ for the amount of capital and labour used for production of variety i in industries Y or Z respectively.

Each final product needs one specific intermediate. For managing the production chain, IT services are used. The administration rights are distributed according to Lemmas 1 and 2 presented in Sect. 3.4.6.

- When final goods and intermediate goods producers are integrated, the administration rights are centralized. An IT service department is integrated into the headquarters of the integrated company.
- While the companies of the production chain are separated legally, the IT service provider remains independent as well.

For each pair of final and intermediate goods, IT services are necessary, which are produced with a fixed amount of capital K_0 and labour L_0, which correspond to the factor endowment of the integrated economy[1]:

$$K_0 = \zeta K^* \tag{5.8}$$

[1] In Sect. 3.4.5, it is argued that the IT service provider independently decides about investments in human capital. Though it is assumed here for the sake of simplicity that qualified and less-qualified labour force is aggregated into L, compensated with average wage w.

$$L_0 = \zeta L^* \tag{5.9}$$

with parameter $0 < \zeta < 1$.

The IT service costs generated by the production of a variant of the final good are:

$$D = rK_0 + wL_0 = r\zeta K^* + w\zeta L^* \tag{5.10}$$

Free entry into the IT service sector is assumed so that profits for IT service companies are zero and D can be integrated into the profit function of the intermediate goods producer. It is assumed that the intermediate suppliers have to cover the costs of all IT services. With regard to these assumptions, it is emphasized in Sect. 5.3 that international trade in IT services occurs without differences in production functions or factor input.

The production function of a variant of the final product of sector Y yields[2]

$$y(i) = x(i) \tag{5.11}$$

Total cost function for producing a variant i of the final product is:

$$C_Y(i) = r(K_Y(i) + K_0) + w(L_Y(i) + L_0) + T - T \tag{5.12}$$

with T for the lump-sum transfer, paid by the intermediate producer (-T) to the final goods producer (+T) at the beginning of their relationship. For sector Z respectively.

5.2.3 *Organization*

5.2.3.1 Timing of the Nash Bargaining

The contracts between final and intermediate goods producers are incomplete in the sense of Grossman and Hart (1986). Two organizational forms between the producers can occur: integration, or separation. The decision on the relationship between the two companies depends on the anticipated result of a Nash bargaining process. The final goods producer invests in capital K and the intermediate producer in labour L. In detail, the timing of events in the negotiations between final and intermediate goods producers are as follows:

- t_0: Choice of organizational structure: integration, or separation. This implies decisions about the ownership of K (or if K is lent to the supplier by the final goods producer) and the volume of ex ante transfer T, which is paid from the supplier to the final goods producer to buy into the relationship.

[2] Alternatively, a production function such as $y(i) = \chi(K_0, L_0)x(i)$ with a efficiency parameter $\chi > 1$ could be assumed. When χ would depend on capital and labour invested in IT services, the model would lose mathematical tractability without gaining further insight.

Table 5.1 Value of outside options (First Number) and share of ex-post gains from trade (Second Number) in the case of a successful Nash-bargaining process

Profit share	Final goods producer	Intermediate goods producer
Incomplete contracts: Integration	$\delta^{\alpha}R_Y{}^* + \phi(1-\delta^{\alpha})R_Y{}^*$ $= \bar{\phi}R_Y{}^*$	$0 + (1-\phi)(1-\delta^{\alpha})R_Y{}^*$ $= (1-\bar{\phi})R_Y{}^*$
Incomplete contracts: Separation	$0 + \phi\, R_Y{}^*$	$0 + (1-\phi)R_Y{}^*$
Complete contracts	$\phi(R_Y{}^* - rK)$	$(1-\phi)(R_Y{}^* - wL)$

- t_1: The contracts with the IT service company and the ex ante investments in capital and labour are carried out.
- t_2: The intermediates X are produced.
- t_3: The producers bargain over the surplus after having sold the final products, the Nash bargaining process takes place. Volumes of ex ante investment and the quality of the intermediates are observable for both companies. The values of the outside options are shown in Table 5.1.
- t_4: The final goods are produced and sold.

First, the parties are remunerated according to their outside options, then the residual revenues are bargained over. The outside options are 0 for the intermediate goods producer either by integration or separation because the intermediates are specific for the final products. For the final goods producer, the outside option is 0 in the case of separation because it has no possibility of changing to another intermediate producer by the assumptions made above. In the case of integration, it is assumed that the final goods producer holds only a fraction $0 < \delta < 1$ of the residual right to use the x(i) for selling final products y(i). The amount of $\delta y(i)$ can be sold for a price which is $p_y(i) = A_y^{1-\alpha}(\delta y(i))^{-(1-\alpha)}$ according to (5.4).

The potential revenues from the sale of a final product of sector Y are:

$$R_Y(i)^* = p_y(i)y(i) \tag{5.13}$$

Rearranging (5.4) leads to:

$$p_y(i) = \left(\frac{y(i)}{A_Y}\right)^{\alpha-1}$$

Including this, (5.13) and (5.7) results in:

$$R_Y(i)^* = A_Y^{1-\alpha}y^{\alpha} = A_Y^{1-\alpha}\left(\frac{K_{x,Y}(i)}{\beta_Y}\right)^{\alpha\beta_Y}\left(\frac{L_{x,Y}(i)}{1-\beta_Y}\right)^{\alpha(1-\beta_Y)} \tag{5.14}$$

By assumption, the bargaining process in t_3 ends up with a share of $1/2 < \phi < 1$ from the residual revenues which are taken by the final goods producers. If the negotiating process has been successful in the case of separation, the revenues for the final goods producers amount ϕ of potential revenues R_Y^*, and $1-\phi$ for the intermediate goods producer respectively. In the case of integration, the final goods producer first gets a amount of $(\delta y)(A_y^{1-\alpha}(\delta y(i))^{-(1-\alpha)})$ which equals a share δ^α of potential revenues R_Y^*. The remaining $(1-\delta^\alpha)R_Y^*$ is shared according to ϕ. Table 5.1 summarizes the revenue sharing.

The table shows that the supplier has weaker bargaining power in the case of integration, and a relatively stronger position in the case of separation. Anticipating the Nash bargaining process, the companies decide on their investment volumes in t_1.

5.2.3.2 Optimal Output

According to Table 5.1, the final goods producers maximize profit by:

$$\overline{\phi} R_Y(i) - rK_{x,Y,I}(i) \Rightarrow MAX! \tag{5.15}$$

with

$$\overline{\phi} = \delta^\alpha + \phi(1-\delta^\alpha) > \phi$$

by choosing the investment volume of $K_{x,Y,I}(i)$.

Simultaneously, the integrated supplier maximizes its profit function:

$$(1-\overline{\phi})R_Y(i) - wL_{x,Y,I}(i) - D \Rightarrow MAX! \tag{5.16}$$

Analogously, in the case of separation, the companies maximize:

$$\phi R_Y(i) - rK_{x,Y,S}(i) - D \Rightarrow MAX! \tag{5.17}$$

and:

$$\phi R_Y(i) - wL_{x,Y,S}(i) - D \Rightarrow MAX! \tag{5.18}$$

by choosing the investment volumes of $K_{x,Y,S}(i)$ and $L_{x,Y,S}(i)$ respectively.

Optimization

By inserting variable ψ_l with $l \in I, S$ and $\psi_I = \overline{\phi}$ and $\psi_S = \phi$ the maximization procedure can be calculated for both cases, integration and separation, simultaneously. The summands with D vanish with differentiation. For details of the calculation, see Appendix 12.6.

Maximizing (5.15) and (5.16) and solving the two resulting response-functions for $L_{x,Y,l}(i)$ and $K_{x,Y,l}(i)$ show the known factor-input relationship of Cobb-Douglas production functions:

$$\frac{\psi_l}{1-\psi_l}\frac{w}{r}\frac{L_{Y,l}}{1-\beta_Y} = \frac{K_{Y,l}}{\beta_Y} \tag{5.19}$$

Please note that (5.19) does not show the total factor-input ratio of the sector because the costs for IT services are not in the first derivatives of the profit functions. IT service costs rK_0 and rL_0 are sufficiently small by assumption, so that positive yields are realized. Solving the maximization problem, the equilibrium ex ante investments yield:

$$K_{x,Y,l}(i) = \beta_Y A_Y \left({\psi_l}^{1-\alpha(1-\beta_Y)} (1-\psi_l)^{\alpha(1-\beta_Y)} r^{\alpha(1-\beta_Y)-1} w^{-\alpha(1-\beta_Y)} \alpha \right)^{\frac{1}{1-\alpha}} \tag{5.20}$$

and:

$$L_{x,Y,l}(i) = A_Y(1-\beta_Y) \left(\alpha \psi_l^{\alpha\beta_Y} (1-\psi_l)^{1-\alpha\beta_Y} w^{\alpha\beta_Y - 1} r^{-\alpha\beta_Y} \right)^{\frac{1}{1-\alpha}} \tag{5.21}$$

(5.20) and (5.21) into (5.11) and (5.7) gives optimal output and prices:

$$\begin{aligned} y_l = A_Y & \left({\psi_l}^{1-\alpha(1-\beta_Y)} (1-\psi_l)^{\alpha(1-\beta_Y)} r^{\alpha(1-\beta_Y)-1} w^{-\alpha(1-\beta_Y)} \alpha \right)^{\frac{\beta_Y}{1-\alpha}} \\ & \times \left(\alpha \psi_l^{\alpha\beta_Y} (1-\psi_l)^{1-\alpha\beta_Y} w^{\alpha\beta_Y - 1} r^{-\alpha\beta_Y} \right)^{\frac{1-\beta_Y}{1-\alpha}} \end{aligned}$$

Rearranging leads to:

$$y_l = A_Y \left(\alpha r^{-\beta_Y} w^{\beta_Y - 1} \psi_l^{\beta_Y} (1-\psi_l)^{1-\beta_Y} \right)^{\frac{1}{1-\alpha}} \tag{5.22}$$

Solving (5.4) for p_Y and inserting in: (5.22) lead to:

$$p_{Y,l} = \frac{1}{\alpha r^{-\beta_Y} w^{\beta_Y - 1} \psi_{lY}^{\beta} (1-\psi_l)^{1-\beta_Y}} = \frac{r_Y^{\beta} w^{1-\beta_Y}}{\alpha \psi_l^{\beta_Y} (1-\psi_l)^{1-\beta_Y}} \tag{5.23}$$

Marginal costs for producing an additional unit of an industry y-variant are $r_Y^{\beta} w^{1-\beta_Y}$. The standard result of models with monopolistic competition is that the price is $1/\alpha$ above marginal costs. Equation (5.23) shows that the price is $\psi_l^{\beta_Y}(1-\psi_l)^{1-\beta_Y}$ times higher than in a scenario with complete contracts (see Sect. 5.2.3):the incomplete contracts lead to distortions. The lower β_Y the more important is labour and the higher is the mark-up.

The relation between ex ante equilibrium revenues and factor costs is constant. For the final goods producer it is:

$$\frac{p_{Y,l} y_l}{r K_{Y,l}} = \frac{A \left(\alpha r^{-\beta_Y} w^{\beta_Y - 1} \psi_l^{\beta_Y} (1-\psi_l)^{1-\beta_Y} \right)^{\frac{\alpha}{1-\alpha}}}{r \beta_Y A_Y \left({\psi_l}^{1-\alpha(1-\beta_Y)} (1-\psi_l)^{\alpha(1-\beta_Y)} r^{\alpha(1-\beta_Y)-1} w^{-\alpha(1-\beta_Y)} \alpha \right)^{\frac{1}{1-\alpha}}}$$

$$\frac{p_{Y,l}\, y_l}{rK_{Y,l}} = \frac{1}{\alpha\beta_Y\psi} \tag{5.24}$$

Similarly, the result for the supplier is:

$$\frac{p_{Y,l}\, y_l}{wL_{Y,l}} = \frac{1}{\alpha(1-\beta_Y)(1-\psi)} \tag{5.25}$$

Profit Functions

Now the expected profit of the supplier can be calculated. Due to competition between the suppliers, the expected profit equals the lump-sum transfer T which is used in t_0 to buy into the relationship so that total profits of the supplier are zero in general equilibrium. Undoing the simplifying notification with $\psi_I = \overline{\phi}$ and $\psi_S = \phi$ the lump-sum transfer in the case of integration yields according to (5.16):

$$T_{Y,I} = (1-\overline{\phi})R_Y - wL_{x,Y} - D$$

$$T_{Y,I} = (1-\overline{\phi})p_{Y,l}y_l - wL_{Y,l} - D$$

Using (5.25) and (5.4) leads to:

$$T_{Y,I} = (1-\overline{\phi})(1-\alpha(1-\beta))A_Y\, p_{Y,l}^{\frac{-\alpha}{1-\alpha}} - D \tag{5.26}$$

Similarly, the lumpsum transfer in the case of separation is determined:

$$T_{Y,S} = (1-\phi)R_Y - wL_{x,Y} - D$$

$$T_{Y,S} = (1-\phi)(1-\alpha(1-\beta))A_Y\, p_{Y,l}^{\frac{-\alpha}{1-\alpha}} - D \tag{5.27}$$

Using (5.15) and (5.17) now the ex ante profit function of the final goods producer can be calculated. For the case of integration:

$$\Pi_{F,Y,I} = \overline{\phi}p_{Y,I}y_I - rK_{Y,I} + T_{Y,I} - D$$

And for the case of separation:

$$\Pi_{F,Y,S} = \phi p_{Y,S}y_S - rK_{Y,S} + T_{Y,S} - 0,5D$$

After rearranging and using (5.4) and (5.24) this leads for the case of **integration** to:

$$\Pi_{F,Y,I} = (\overline{\phi} - \overline{\phi}\alpha\beta + (1-\overline{\phi})(1-\alpha+\alpha\beta))A_Y\, p_{Y,I}^{\frac{-\alpha}{1-\alpha}} - D$$

$$\Pi_{F,Y,I} = (1-\alpha+\alpha\beta-2\alpha\beta\overline{\phi}+\alpha\overline{\phi})A_Y\, p_{Y,I}^{\frac{-\alpha}{1-\alpha}} - D \tag{5.28}$$

with

$$p_{Y,I} = \frac{r^{\beta} w^{1-\beta}}{\alpha \overline{\phi}^{\beta} (1 - \overline{\phi})^{1-\beta}}$$

and for the case of **separation** to the profit function of the final goods producer of:

$$\Pi_{F,Y,S} = (\phi - \phi\alpha\beta + (1-\phi)(1-\alpha+\alpha\beta)) A_Y \, p_{Y,S}^{\frac{-\alpha}{1-\alpha}} - D$$

$$\Pi_{F,Y,S} = (1 - \alpha + \alpha\beta - 2\alpha\beta\phi + \alpha\phi) A_Y \, p_{Y,S}^{\frac{-\alpha}{1-\alpha}} - D \tag{5.29}$$

with

$$p_{Y,S} = \frac{r^{\beta} w^{1-\beta}}{\alpha \phi^{\beta} (1 - \phi)^{1-\beta}}.$$

Equations (5.28) and (5.29) show that on a more aggregate level it makes no difference in financial terms who finances the IT service costs in the first instance. The share of the IT service costs paid by the supplier are carried forward.

5.2.3.3 Comparison of Separation, Integration and Complete Contracts

In the case of complete contracts, intermediate and final goods producers share the total profit $p_y(i)y(i) - rK_Y(i) - wL_Y(i) - D$ according to the negotiated shares ϕ and $(1-\phi)$. Figure 5.1 also shows the differences. The two companies optimize their profits by choosing K (final product producer) and L (intermediate product producer), assuming that the fixed costs for IT services are small enough so that profit remains positive. Optimization (see Appendix 12.7 for details) leads to the factor demands of:

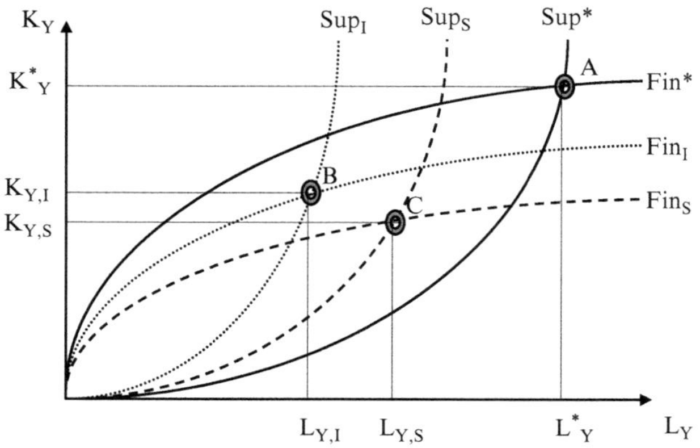

Fig. 5.1 Complete vs. incomplete contracts with $\overline{\phi} > \phi > 1/2$; adapted from Antras (2003), p. 1389

$$L_Y^c = A_Y(1-\beta)\left(\alpha w^{\alpha\beta-1} r^{-\alpha\beta}\right)^{\frac{1}{1-\alpha}} \tag{5.30}$$

and

$$K_Y^c = \beta A_Y\left(r^{\alpha(1-\beta)-1} w^{-\alpha(1-\beta)}\alpha\right)^{\frac{1}{1-\alpha}} \tag{5.31}$$

Comparing (5.30) with (5.21) ánd (5.31) with (5.20) shows for $0 < \psi < 1$ that $K_Y^c > max\{K_{Y,I}, K_{Y,S}\}$ and $L_Y^c > max\{L_{Y,I}, L_{Y,S}\}$. This formally proves the hold-up problem, that in the case of incomplete contracts investments are lower than in the case of complete contracts. In the case of complete contracts, companies can be sure that they are fully rewarded for their investments. This is not the case with incomplete contracts, therefore, their investments in capital and labour are lower depending on their negotiating power. In Fig. 5.1, the point A showing the case of complete contracts is in the upper corner of the K-L diagram.

The curves Sup and Fin represent the reaction functions of the supplier and the final goods producer in the scenarios of complete contracts (*), integration (I), or separation (S). The curves are a result of the optimization process described before. The investments are more capital-intensive in the case of integration (point B), and more labour-intensive in the case of separation (point C). This directly follows from Table 5.1, because the intermediate goods producer has a relatively stronger bargaining position in the case of separation.

Formally, the relative distance between the curves Fin_I, Fin_S remains constant, curves Sup_I, Sup_S respectively. For example, the ratio between the demand of capital from an integrated final goods producer and a separated final goods producer according to (5.20) yields:

$$\frac{K_{Y,I}}{K_{Y,S}} = \frac{\beta_Y A_Y\left(\overline{\phi}_l^{\,1-\alpha(1-\beta)}(1-\overline{\phi}_l)^{\alpha(1-\beta_Y)} r^{\alpha(1-\beta_Y)-1} w^{-\alpha(1-\beta_Y)}\alpha\right)^{\frac{1}{1-\alpha}}}{\beta_Y A_Y\left(\phi_l^{\,1-\alpha(1-\beta)}(1-\phi_l)^{\alpha(1-\beta_Y)} r^{\alpha(1-\beta_Y)-1} w^{-\alpha(1-\beta_Y)}\alpha\right)^{\frac{1}{1-\alpha}}}$$

which reduces to:

$$\frac{K_{Y,I}}{K_{Y,S}} = \left[\frac{\overline{\phi}}{\phi}\frac{1-\phi}{1-\overline{\phi}}\right]^{\frac{1-\alpha(1-\beta_Y)}{1-\alpha}} > 1$$

for the case of $\overline{\phi} > \phi > 1/2$. According to its reaction function, the final goods producer in an integrated company always chooses a higher investment in capital due to a stronger negotiating position. The reasoning behind the other relationships shown in Fig. 5.1 can be developed accordingly: a supplier will choose a higher investment in labour in the case of separation. In other words, the supplier would choose a higher capital-labour ratio in the case of integration, which leads to the industry-specific investment behaviour described. Section 5.2.5 will use this in the context of the general equilibrium.

5.2.4 Ownership Structures

The profit situation of a final goods producer also depends on the organizational structure, as the comparison between (5.28) and (5.29) shows. The final goods producer will integrate its supplier, if:

$$\Pi_{F,Y,I} > \Pi_{F,Y,S}$$

Subtracting the costs for IT services on both sides leads to the condition for integration that:

$$\theta = \frac{\Pi_{F,Y,I} - D}{\Pi_{F,Y,S} - D} > 1 \tag{5.32}$$

Using the relationship $\overline{\phi} = \delta^{\alpha} + \phi(1 - \delta^{\alpha})$ and (5.28) and (5.29), the parameter θ can be expressed as a function of the model's parameters:

$$\theta = \frac{1 - \alpha + \alpha\beta - (2\alpha\beta - \alpha)(\delta^{\alpha} + \phi(1 - \delta^{\alpha}))}{1 - \alpha + \alpha\beta - (2\alpha\beta - \alpha)\phi} \times \left(\frac{\phi^{\beta}(1 - \phi)^{1-\beta}}{[\delta^{\alpha} + \phi(1 - \delta^{\alpha})]^{\beta}(1 - (\delta^{\alpha} + \phi(1 - \delta^{\alpha})))^{1-\beta}}\right)^{\frac{-\alpha}{1-\alpha}}$$

Rearranging with $1 - (\delta^{\alpha} + \phi(1 - \delta^{\alpha})) = (1 - \delta^{\alpha})(1 - \phi)$ leads to:

$$\theta = \left(1 + \frac{\alpha(2\beta - 1)\delta^{\alpha}(\phi - 1)}{1 - \alpha + \alpha\beta - (2\alpha\beta - \alpha)\phi}\right)\left(\frac{\phi^{\beta}(1 - \delta^{\alpha})^{\beta}}{[\delta^{\alpha} + \phi(1 - \delta^{\alpha})]^{\beta}(1 - \delta^{\alpha})}\right)^{\frac{-\alpha}{1-\alpha}}$$

$$\theta = \left(1 + \frac{\alpha(2\beta - 1)\delta^{\alpha}(\phi - 1)}{1 - \alpha + \alpha\beta - (2\alpha\beta - \alpha)\phi}\right)\left(1 + \frac{\delta^{\alpha}}{\phi(1 - \delta^{\alpha})}\right)^{\frac{\alpha\beta}{1-\alpha}}(1 - \delta^{\alpha})^{\frac{\alpha}{1-\alpha}} \tag{5.33}$$

The organizational structure between final and intermediate goods producers depends on the fundamental parameters of the model. For $\theta > 1$, all suppliers will be integrated, for $\theta < 1$, they will remain independent.

For further analysis, the elasticity β is focussed upon. The derivate yields:

$$\frac{\partial\theta}{\partial\beta} > 0$$

for all $0 < \beta < 1$. For details, see Appendix 12.7.1. The attractiveness of integration increases with the capital intensity of the intermediate production, because an underinvestment in a capital-intensive sector harms more than capital underinvestment in a labour-intensive one.

Further, it is assumed that there is a knife-edge case[3] $\widehat{\beta}$ for which $\theta = 1$. It can be differentiated between three cases:

- Pervasive integration: $\beta > \widehat{\beta}$;
- Mixed integration (knife-edge case): $\beta = \widehat{\beta}$; and
- Pervasive outsourcing (separation): $\beta < \widehat{\beta}$.

To rebuild an economy with diversified organizational structures, it is further assumed that one industry is more capital-intensive and the other is more labour-intensive than a reference industry with elasticity $\widehat{\beta}$:

$$\beta_Z < \widehat{\beta} < \beta_Y \tag{5.34}$$

which means that industry Y is relatively more capital-intensive and the final and intermediate goods producers tend to integrate. Please note that the contract between IT service companies and the goods producers follows the organizational structure of the main contract as described in Sect. 3.4.

5.2.5 General Equilibrium

With these elements, the model for the closed economy can be described completely by calculation of the number of companies, sector-specific output, the demand for labour and capital, and the wage-interest rate ratio. Each of these sector-specific variables vary with the organizational structure chosen. The knife-edge case with $\beta = \widehat{\beta}$ is not considered further.

In a general equilibrium, total expenditure equals factor revenues $E = rK^* + wL^*$ (see (5.6)). The aim now is to calculate the sector-specific factor demands and the output in general equilibrium, which strongly depends on the number of companies n.

Integration

First, the number of companies in the integrated sector Y is calculated. In equilibrium, free entry implies that profit as defined in (5.28) becomes zero. According to standard results in models with monopolistic competition, the contribution margins are used to cover fixed costs. So (5.28) becomes:

$$A_Y p_{Y,I}^{\frac{-\alpha}{1-\alpha}} = \frac{D}{1 - \alpha + \alpha\beta_Y - 2\alpha\beta_Y\overline{\phi} + \alpha\overline{\phi}} \tag{5.35}$$

[3] The existence of such a $\widehat{\beta}$ can be proved formally. Given the derivative $\frac{\partial\theta}{\partial\beta} > 0$ for all $0 < \beta < 1$ it can be shown that there exists a β_1 for which $\theta < 1$ and a β_2 for which $\theta > 1$. See Antras (2003).

On the other hand, the overall demand for variants of the sector determines the factor $Ap_{Y,I}$.

The firms of a sector behave symmetrically. Here, all companies integrate ($n_{Y,S} = 0$) and charge identical prices p_Y. With this information, (5.5) changes to

$$A_Y = \frac{\mu E}{n_{Y,I}\, p_{Y,I}^{\frac{-\alpha}{1-\alpha}}}$$

Inserting $A_Y\, p_{Y,I}^{\frac{-\alpha}{1-\alpha}}$ into (5.35) leads to the number of integrated firms in sector y:

$$n_Y = \frac{1-\alpha+\alpha\beta_Y - 2\alpha\beta_Y\overline{\phi} + \alpha\overline{\phi}}{D}\mu E = \frac{\mu}{\zeta}(1-\alpha+\alpha\beta_Y-2\alpha\beta_Y\overline{\phi}+\alpha\overline{\phi}) \quad (5.36)$$

The number of companies in industry Y is independent of the endowment with labour and capital: mathematically, demand-side effects of a rise of capital or labour (higher expenditures) μE are offset by higher fixed costs D so that the number of variants remains constant.

With (5.22) and $\psi = \overline{\phi}$, the optimal output of a variant is defined. This can be rearranged to:

$$y_I = \frac{\alpha D}{1-\alpha+\alpha\beta_Y - 2\alpha\beta_Y\overline{\phi}+\alpha\overline{\phi}} \frac{(1-\overline{\phi})^{1-\beta_Y}\overline{\phi}_Y^{\beta}}{w^{1-\beta_Y} r_Y^{\beta}} \quad (5.37)$$

Finally, w and r have to be determined. The wage w is the equilibrium price on the labour market, so that labour demand L^d equals supply L^* with:

$$L^d = L_Y^d + L_Z^d \quad (5.38)$$

Total labour demand from industry Y is given by n-times the labour demand induced by the production process of a variant:

$$L_Y^d = n_Y(L_{x,Y,I} + L_0) \quad (5.39)$$

Using (5.9), (5.21) and (5.36) leads to:

$$L_Y^d = \frac{\mu}{\zeta}(1-\alpha+\alpha\beta_Y - 2\alpha\beta_Y\overline{\phi}+\alpha\overline{\phi})$$
$$\left[A_Y(1-\beta_Y)\left(\alpha\overline{\phi}^{\alpha\beta_Y}(1-\overline{\phi})^{1-\alpha\beta_Y} w^{\alpha\beta_Y-1} r^{-\alpha\beta_Y}\right)^{\frac{1}{1-\alpha}} + \zeta L^*\right]$$

Inserting (5.35) and (5.23) for A, after rearranging this leads to:

$$L_Y^d = \mu(rK^* + wL^*)\frac{\alpha(1-\beta_Y)(1-\overline{\phi})}{w} + \mu L^*(1-\alpha+\alpha\beta_Y - 2\alpha\beta_Y\overline{\phi} + \alpha\overline{\phi}) \tag{5.40}$$

or:

$$L_Y^d = \mu\left[rK^*\frac{\alpha(1-\beta_Y)(1-\overline{\phi})}{w} + L^*(1-\alpha\beta_Y\overline{\phi})\right] \tag{5.41}$$

The first summand shows the effects of a rising endowment with capital: due to a constant number of variants, the output of one variant rises, causing a higher demand of labour with dependence on elasticity of substitution, production technology, and negotiation position of capital owners, which is offset by a rise of the wage w. The second summand shows that a rising endowment with labour has the effect that w has to fall, but this effect is dampened by $\alpha\beta\overline{\phi}$. If ϕ, and, hence, $\overline{\phi}$ are relatively small, the distribution of the profits favours the intermediate department of the integrated company (see Table 5.1) so that they are interested in investing more in labour. The greater ϕ, the worse the underinvestment in labour compared to a world with complete contracts.

The solution for capital is analogue. The demand for capital is:

$$K_Y^d = K_Y^d + K_Z^d \tag{5.42}$$

with

$$K_Y^d = n_Y(K_{x,Y,I} + K_0) \tag{5.43}$$

Inserting (5.20), (5.35) and (5.23) in (5.43) gives:

$$K_Y^d = \frac{\mu}{\zeta}\left[\alpha\beta_Y D\frac{\overline{\phi}}{r} + \zeta K^*(1-\alpha+\alpha\beta_Y - 2\alpha\beta_Y\overline{\phi} + \alpha\overline{\phi})\right]$$

$$K_Y^d = \mu(rK^* + wL^*)\alpha\beta_Y\frac{\overline{\phi}}{r} + \mu K^*(1-\alpha+\alpha\beta_Y - 2\alpha\beta_Y\overline{\phi} + \alpha\overline{\phi}) \tag{5.44}$$

Rearranging further leads to:

$$K_Y^d = \mu\left[\alpha\beta_Y\overline{\phi}\frac{w}{r}L^* + K^*(1-\alpha(1-\overline{\phi}) + \alpha\beta_Y(1-\overline{\phi}))\right] \tag{5.45}$$

The higher ϕ, the higher are $\overline{\phi}$ and the demand for capital. In the second summand $0 < \alpha\beta_Y < \alpha < 1$: when $\overline{\phi}$ rises, the diminishing effect of $\alpha\beta_Y\phi$ is outweighed by $\alpha\phi$.

Separation

For sector Z, which is characterized by the separation of final, intermediate goods producers and IT service companies, the arguments are similar. Equation (5.5) for sector Z reads:

$$A_Z = \frac{(1-\mu)E}{\int_0^{n_{Z,I}} p_{Z,I}(j)^{-\frac{\alpha}{1-\alpha}}dj + \int_0^{n_{Z,S}} p_{Z,S}(j)^{-\frac{\alpha}{1-\alpha}}dj} \tag{5.46}$$

With $n_{Z,I} = 0$ and profit function (5.29) set to zero, the optimal number of final products suppliers in this sector is quite similar to (5.36):

$$n_Z = (1-\alpha+\alpha\beta_Z - 2\alpha\beta_Z\phi + \alpha\phi)\frac{(1-\mu)}{\zeta} \tag{5.47}$$

Accordingly, the values for parameter A, variant output z_S, labour and capital demand of sector Z are:

$$A_Z p_{Z,S}^{\frac{-\alpha}{1-\alpha}} = \frac{D}{1-\alpha+\alpha\beta_Z - 2\alpha\beta_Z\phi + \alpha\phi} \tag{5.48}$$

$$z_S = \frac{\alpha D}{1-\alpha+\alpha\beta_Z - 2\alpha\beta_Z\phi + \alpha\phi}\frac{(1-\phi)^{1-\beta_Z}\phi_Z^{\beta}}{w^{1-\beta_Z}r_Z^{\beta}} \tag{5.49}$$

$$L_Z^d = (1-\mu)(rK^*+wL^*)\frac{\alpha(1-\beta_Z)(1-\phi)}{w} + (1-\mu)L^*(1-\alpha+\alpha\beta_Z-2\alpha\beta_Z\phi+\alpha\phi) \tag{5.50}$$

or:

$$L_Z^d = (1-\mu)\left[rK^*\frac{\alpha(1-\beta_Z)(1-\phi)}{w} + L^*(1-\alpha\beta_Z\phi)\right] \tag{5.51}$$

$$K_Z^d = (1-\mu)(rK^*+wL^*)\alpha\beta_Z\frac{\phi}{r} + (1-\mu)K^*(1-\alpha+\alpha\beta_Z-2\alpha\beta_Z\phi+\alpha\phi) \tag{5.52}$$

or

$$K_Z^d = (1-\mu)\left[\alpha\beta_Z\phi\frac{w}{r}L^* + K^*(1-\alpha(1-\phi)+\alpha\beta_Z(1-\phi))\right] \tag{5.53}$$

Comparing the equations for industries Y and Z shows that the differences between the sectors are the factor intensities $\beta_Z < \beta_Y$ and the different negotiation position, as characterized by the difference between ϕ and $\overline{\phi}$. Nevertheless, also in the case of separation, a high ϕ favours capital holders, a small ϕ favours labour.

Now, the wage-interest ratio can be calculated. In equilibrium, the factor prices adjust so that labour demand equals labour supply. Inserting (5.41) and (5.51) in (5.38) yields[4]:

$$\frac{w}{r} = \frac{K^*[(1-\beta_Y)(1-\overline{\phi}) + (1-\beta_Z)(1-\phi)]}{L^*[\beta_Y\overline{\phi} + \beta_Z\phi]} \tag{5.54}$$

[4] Alternatively, one factor can be chosen as numeraire for the system, so that w and r could be calculated unambiguously.

This equation has a striking implication for the distribution of income: if the negotiation power of the final goods producer against their supplier rises (higher values of ϕ and $\overline{\phi}$), the wage-interest rate ratio falls. Interpreting the assumptions for real-world situations, the equations imply that in a case of capital-intensive holdings as final goods producers and labour-intensive SMEs as suppliers, the best policy for labour unions would be to support the negotiating power of SME.

Finally, the distribution of income can be calculated directly:

$$\frac{L^*w}{K^*r} = \frac{(1-\beta_Y)(1-\overline{\phi}) + (1-\beta_Z)(1-\phi)}{\beta_Y\overline{\phi} + \beta_Z\phi} \tag{5.55}$$

5.3 Two-Country Model

5.3.1 *Factor Price Equalization*

The aim of the model is to describe the exchange pattern of international IT services. Therefore, the integrated world presented in the previous section is split into two countries, A and B. To stress the effects of international trade, some further assumptions are made: preferences and production technologies are identical in both countries, but endowments in capital and labour differ. These are the standard assumptions of Heckscher-Ohlin-type models, but here the final products as well as the input factors are assumed to be nontradable. Only intermediate products and associated IT services are tradable, either in-house (integrated companies), or between companies (outsourcing solution). Following the approach of Helpman and Krugman (1985), it is assumed that factor price equalization holds, i.e. that the capital-labour ratio between the two countries does not differ 'too much', so factor price equalization is caused by trade in the intermediate goods.

Due to the fact that the optimal output of an intermediate is also determined by the costs of IT services and the mark-up parameter of the model (monopolistic competition) which are identical in both countries, the pivotal parameter for the volume of trade is the number n of variants produced. The next step, therefore, is to determine the equilibrium numbers of variants of industries Y and Z produced in countries A and B.

The number of variants of each country sum up to number of variants in an integrated world economy:

$$n_Y^* = n_Y^A + n_Y^B \tag{5.56}$$

$$n_Z^* = n_Z^A + n_Z^B \tag{5.57}$$

Having determined overall demand for labour and capital, the country-specific demand for both factors depends on the share of variants produced in the country. Adding up (5.52) and (5.44) by using the country-specific weights n_Y^A/n_Y^* and

n_Z^A/n_Z^* gives the demand for capital and labour of country A:

$$K^A = \frac{n_Y^A}{n_Y^*}\left[\mu(rK^* + wL^*)\alpha\beta_Y\frac{\overline{\phi}}{r} + \mu K^*(1-\alpha+\alpha\beta_Y - 2\alpha\beta_Y\overline{\phi}+\alpha\overline{\phi})\right]$$
$$+\frac{n_Z^A}{n_Z^*}\left[(1-\mu)(rK^* + wL^*)\alpha\beta_Z\frac{\phi}{r}\right.$$
$$\left. + (1-\mu)K^*(1-\alpha+\alpha\beta_Z - 2\alpha\beta_Z\phi+\alpha\phi)\right] \tag{5.58}$$

From (5.40) and (5.50)

$$L^A = \frac{n_Y^A}{n_Y^*}\left[\mu(rK^* + wL^*)\frac{\alpha(1-\beta_Y)(1-\overline{\phi})}{w}\right.$$
$$\left.+\mu L^*(1-\alpha+\alpha\beta_Y - 2\alpha\beta_Y\overline{\phi}+\alpha\overline{\phi})\right]$$
$$+\frac{n_Z^A}{n_Z^*}\left[(1-\mu)(rK^* + wL^*)\frac{\alpha(1-\beta_Z)(1-\phi)}{w}\right.$$
$$\left.+(1-\mu)L^*(1-\alpha+\alpha\beta_Z - 2\alpha\beta_Z\phi+\alpha\phi)\right] \tag{5.59}$$

Solving this system for $\frac{n_Y^A}{n_Y^*}$ and $\frac{n_Z^A}{n_Z^*}$ yields:

$$\frac{n_Y^A}{n_Y^*} = \frac{\left(\frac{L^A}{L^*}\left[\xi\alpha\beta_Z\phi + \varepsilon_Z\right] - \frac{K^A}{K^*}\left[\frac{\xi}{\xi-1}\alpha(1-\beta_Z)(1-\phi)+\varepsilon_Z\right]\right)}{\mu\left[DEN1 - DEN2\right]} \tag{5.60}$$

$$\frac{n_Z^A}{n_Z^*} = \frac{\frac{1}{(1-\mu)}\left(\frac{L^A}{L^*}\left[\xi\alpha\beta_Y\overline{\phi} + \varepsilon_Y\right] - \frac{K^A}{K^*}\left[\frac{\xi}{\xi-1}\alpha(1-\beta_Y)(1-\overline{\phi})+\varepsilon_Y\right]\right)}{DEN2 - DEN1} \tag{5.61}$$

with:

$$\begin{aligned}
\xi &= (1+\frac{wL^*}{rK^*})\\
\frac{\xi}{\xi-1} &= (1+\frac{rK^*}{wL^*})\\
\varepsilon_Z &= 1-\alpha+\alpha\beta_Z-2\alpha\beta_Z\phi+\alpha\phi\\
\varepsilon_Y &= 1-\alpha+\alpha\beta_Y-2\alpha\beta_Y\overline{\phi}+\alpha\overline{\phi}\\
DEN1 &= \left(\frac{\xi}{\xi-1}\alpha(1-\beta_Y)(1-\overline{\phi})+\varepsilon_Y\right)\left[\xi\alpha\beta_Z\phi+\varepsilon_Z\right]\\
DEN2 &= \left(\frac{\xi}{\xi-1}\alpha(1-\beta_Z)(1-\phi)+\varepsilon_Z\right)\left[\xi\alpha\beta_Y\overline{\phi}+\varepsilon_Y\right]
\end{aligned}$$

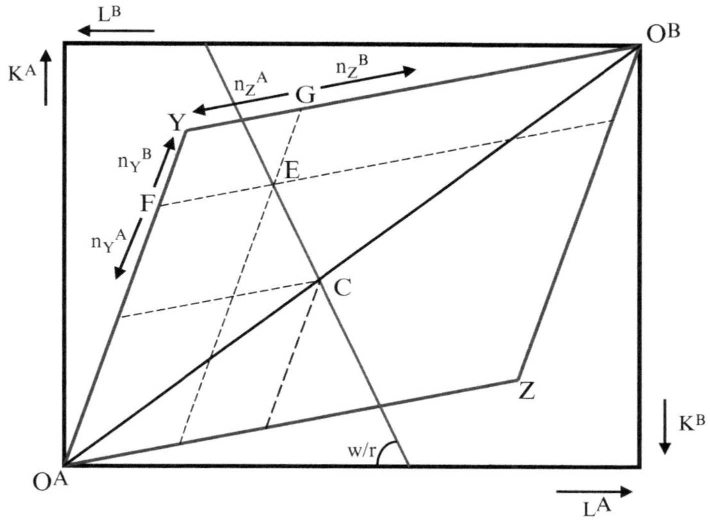

Fig. 5.2 Factor price equalization region; adapted from Antras (2003), p. 1397

For details of the calculations, please see Appendix 12.7.2.

Figure 5.2 shows the factor price equalization region of the two-country model. World endowment with capital and labour is represented by the total size of the box. Point E represents the division of the factors to country A, as measured from origin O^A, and country B with origin O^B. With factor price equalization, the slope of the line EC represents the wage-interest rate ratio. The lines of the parallelogram $O^A YO^B Z$ show the factor intensities producing goods in industry Y (line $O^A Y$) and Z (line $O^A Z$). In these lines, the necessary factor inputs for IT services are already incorporated. The factor intensity of the IT service sector itself is represented by line $O^A O^B$. Variants of industry Y are produced in a capital-intensive manner; variants of industry Z are produced with a more labour-intensive fashion.

Shifting the lines of the parallelogram through point E leads to points F and G, which represent the number of variants produced by countries A and B. In an equilibrium, country A produces $n_Y^A = |O^A F|$ variants of sector Y and $n_Z^A = |YG|$ variants of sector Z measured in units of labour and capital employed.[5] Analogously, country B produces $n_Y^B = |FY|$ variants of sector Y and $n_Z^B = |O^B G|$ variants of sector Z.

Factor price equalization here is not driven by trade in factors or final products, but by trade in intermediates. All factor allocations inside the parallelogram

[5] The basic concept of constructing the parallelogram is using the unit value isoquants (see Helpman and Krugman (1985)). So a constant length of one side of the parallelogram could represent a higher output when the price for the product lowers, e.g. production becomes cheaper.

$O^A YO^B Z$, such as E, allow for factor price equalization so that both countries produce intermediates for both industries. If E was on the outer line of $O^A YO^B Z$, there would be factor price equalization, but one country would specialize fully in intermediates of the industry.

To assume that factor price equalization holds, it is necessary that $n_Y^j > 0$ and $n_Z^j > 0$ for both countries $\{A, B\}$. In the following, country A is considered. To explore the conditions $\frac{n_Y^A}{n_Y^*} > 0$ *and* $\frac{n_Z^A}{n_Z^*} > 0$ the symmetry of the above equations (the denominator of (5.60) equals the denominator of (5.61) times (-1) and the factor μ) is used. Excluding the case that the denominator becomes zero, in the following, the case with a negative denominator of (5.60) and a positive denominator of (5.61) is considered here. (The other case is omitted because it is contradictory to the basic assumptions of the model (see next fnn.)). Mathematically, this assumption is expressed as $DEN2 > DEN1$ or

$$\frac{\xi\alpha\beta_Y\overline{\phi} + \varepsilon_Y}{\frac{\xi}{\xi-1}\alpha(1-\beta_Y)(1-\overline{\phi}) + \varepsilon_Y} > \frac{\xi\alpha\beta_Z\phi + \varepsilon_Z}{\frac{\xi}{\xi-1}\alpha(1-\beta_Z)(1-\phi) + \varepsilon_Z} \tag{5.62}$$

Equation (5.62) holds when, for the left hand side:

$$\xi\alpha\beta_Y\overline{\phi} + \varepsilon_Y > \frac{\xi}{\xi-1}\alpha(1-\beta_Y)(1-\overline{\phi}) + \varepsilon_Y$$

and the right hand side:

$$\xi\alpha\beta_Z\phi + \varepsilon_Z < \frac{\xi}{\xi-1}\alpha(1-\beta_Z)(1-\phi) + \varepsilon_Z$$

Using (5.62) for $\xi - 1$ this can be transformed to:

$$\frac{\beta_Y\overline{\phi}}{(1-\beta_Y)(1-\overline{\phi})} > \frac{rK^*}{wL^*}$$

and:

$$\frac{rK^*}{wL^*} > \frac{\beta_Z\phi}{(1-\beta_Z)(1-\phi)}$$

Putting both inequalities together leads to the condition:

$$\frac{\beta_Y\overline{\phi}}{(1-\beta_Y)(1-\overline{\phi})} > \frac{rK^*}{wL^*}\left[= \frac{\beta_Z\phi + \beta_Y\overline{\phi}}{(1-\beta_Z)(1-\phi) + (1-\beta_Y)(1-\overline{\phi})}\right] > \frac{\beta_Z\phi}{(1-\beta_Z)(1-\phi)} \tag{5.63}$$

The wage-interest rate relationship (5.54) of the integrated world economy, which is valid for country A if factor price equalization occurs, is shown in brackets.

In addition to the assumption made earlier, $\overline{\phi} > \phi > 1/2$, it has to assumed here that the capital intensity of the capital-intensive sector β_Y and the labour intensity of the labour-intensive sector $(1 - \beta_Z)$ must be sufficiently large enough for a solution. Economically, the relative labour intensity of the labour-intensive sector $(1-\beta_Z)/\beta_Z$ must be large enough to outweigh the worse relative negotiation power $\phi/(1-\phi)$ of the suppliers so that (5.63) holds[6].

Furthermore, when (5.62) holds, the conditions for $\frac{n_Z^A}{n_Z^*} > 0$ *and* $\frac{n_Y^A}{n_Y^*} > 0$ are determined by the numerators of (5.60) and (5.61):

$$\frac{L^A}{L^*}\left[\xi\alpha\beta_Z\phi + \varepsilon_Z\right] < \frac{K^A}{K^*}\left[\frac{\xi}{\xi-1}\alpha(1-\beta_Z)(1-\phi) + \varepsilon_Z\right] \tag{5.64}$$

and

$$\frac{L^A}{L^*}\left[\xi\alpha\beta_Y\overline{\phi} + \varepsilon_Y\right] > \frac{K^A}{K^*}\left[\frac{\xi}{\xi-1}\alpha(1-\beta_Y)(1-\overline{\phi}) + \varepsilon_Y\right] \tag{5.65}$$

Rearranging this leads to the condition:

$$\frac{\frac{\xi}{\xi-1}\alpha(1-\beta_Y)(1-\overline{\phi}) + \varepsilon_Y}{\xi\alpha\beta_Y\overline{\phi} + \varepsilon_Y} < \frac{K^*}{L^*}\frac{L^A}{K^A} < \frac{\frac{\xi}{\xi-1}\alpha(1-\beta_Z)(1-\phi) + \varepsilon_Z}{\xi\alpha\beta_Z\phi + \varepsilon_Z} \tag{5.66}$$

This determines the upper and lower limits of the relative factor endowment K^A/L^A which leads to factor price equalization.

To further explore the characteristic of this condition, I introduce:

$$\Omega_Y = \frac{L^A_{high}}{K^A_{low}}\frac{K^*}{L^*} \tag{5.67}$$

$$\Omega_Z = \frac{L^A_{low}}{K^A_{high}}\frac{K^*}{L^*} \tag{5.68}$$

According to (5.66), the actual capital-labour endowment of country A determined by Ω^A must lie between

$$\Omega_Z < \Omega^A < \Omega_Y$$

---

[6] The case of a negative denominator of (5.60) and a positive denominator of (5.61) would lead to

$$\frac{\beta_Y\overline{\phi}}{(1-\beta_Y)(1-\overline{\phi})} < \frac{rK^*}{wL^*} = \frac{\beta_Z\phi + \beta_Y\overline{\phi}}{(1-\beta_Z)(1-\phi) + (1-\beta_Y)(1-\overline{\phi})} < \frac{\beta_Z\phi}{(1-\beta_Z)(1-\phi)}$$

This contradicts the basic assumptions of $\beta_Y > \beta_Z$ and $\overline{\phi} > \phi > 1/2$ and is, therefore, not discussed further.

so that factor price equalization holds. Therefore, the factor endowment of country A lies in the factor price equalization region of Fig. 5.2. Economically, this explains that the capital-labour ratio must not differ 'too much' from the endowment ratio of the world economy. L^A_{high}/K^A_{low} describes the highest possible labour-capital ratio so that intermediates for the capital-intensive good Y are still produced. Ω_Y indicates the highest possible labour-capital ratio relative to world endowment, and Ω_Z describes the lowest possible relative labour-capital ratio so that intermediates of the labour-intensive good are produced. $\Omega_Y > 1$ and $\Omega_Z < 1$ is assumed so that deviations from the worldwide endowment ration L^*/K^* are possible.

Rearranging (5.67) and plugging K^A_{low}/K^* into (5.64) leads to a 'border solution' for sector Y:

$$[\xi\alpha\beta_Z\phi + \varepsilon_Z] = \frac{1}{\Omega_Y}\left[\frac{\xi}{\xi-1}\alpha(1-\beta_Z)(1-\phi) + \varepsilon_Z\right]$$

$$\xi\alpha\beta_Z\phi = \frac{1}{\Omega_Y}\frac{\xi}{\xi-1}\alpha(1-\beta_Z)(1-\phi) + (\frac{1}{\Omega_Y} - 1)\varepsilon_Z$$

Dividing by α and ξ and using (5.62) for $\xi - 1 = wL^*/rK^*$ leads to:

$$\Omega_Y\beta_Z\phi = \frac{rK^*}{wL^*}(1-\beta_Z)(1-\phi) + (1-\Omega_Y)\frac{\varepsilon_Z}{\alpha\xi} \tag{5.69}$$

For sector Z, the result is similar. With (5.67) solving for L^A_{low}/L^* and plugging into (5.65):

$$\Omega_Z(\xi-1)\beta_Y\overline{\phi} + (\Omega_Z - 1)\frac{\varepsilon_Y(\xi-1)}{\xi\alpha} = (1-\beta_Y)(1-\overline{\phi})$$

Using (5.62) for $\xi - 1$

$$\Omega_Z\beta_Y\overline{\phi} + (\Omega_Z - 1)\frac{\varepsilon_Y}{\xi\alpha} = (1-\beta_Y)(1-\overline{\phi})\frac{rK^*}{wL^*} \tag{5.70}$$

To produce variants of sector Z in equilibrium, the factor endowment ratio of country A must not differ from the world endowment ratio more than factor Ω_Z, which equals the inverse of the relative negotiating power of the integrated final goods producer $(1-\overline{\phi})/\overline{\phi}$ weighted by the inverse of the relative capital-intensity of the Y-sector $(1-\beta_Y)/\beta_Y$ and factor income ratios (first summand) after being corrected for an effect which is related to the use of factors by IT services (second summand). The typical opportunity costs argument of Heckscher-Ohlin-type models comes through: the high weighted ***capital absorbing power*** $\beta_Y\overline{\phi}$ of sector Y producers allows for production of sector Z products also when country A is relatively rich of capital, so Ω_Z is low. In this case, the line O^AY in the parallelogram of Fig. 5.2 would be steep, so that the probability grows that the factor endowment point E lies within the factor price equalization region and both goods are produced. If this was not the case, the capital-intensive country would specialize in the capital-intensive intermediates only, and not produce the labour-intensive variants. For sector Y (see (5.69)), these arguments hold vice versa.

5.3.2 *Trade in Intermediates*

Figure 5.3 shows the volumes of international and domestic outsourcing and offshoring (in-house) of trade in intermediates. The EU's definition is used here, which is shown in Table 2.9.

For example, the line $|O^A Y|$ shows the total production of variants of sector Y, of which $|O^A F|$ are produced in country A. Due to monopolistic competition (love of variety), consumers in both countries like to consume final products based on country A's intermediates. Please note that the final products cannot be traded themselves by assumption, meaning that the according final goods producers are located in each country.

The distribution of the variants equals the distribution of total expenditure represented by $|O^A C|$ in relation to $|CO^B|$. Basic geometry ($|HC|$ is parallel to $|O^B F|$)

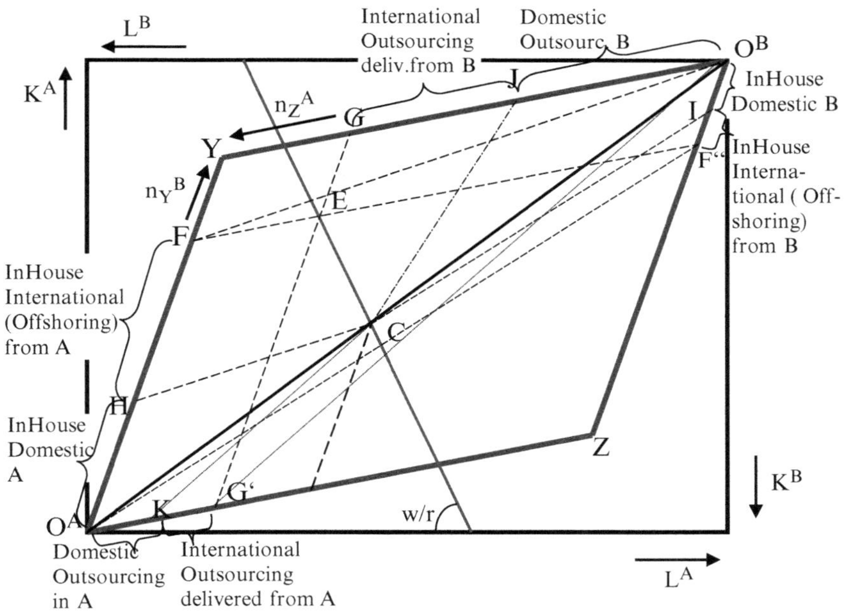

Fig. 5.3 Trade in intermediates

which leads to point H. $|O^A H|$ represents the share of variants which are used domestically.[7]

Accordingly, the other points in Fig. 5.3 are constructed. $|O^A K|$ represents the intermediates which are used domestically and $|KG'|$ the intermediates which are offered to final goods producers in other countries. Outsourcing to A is shown by $|GJ|$, due to country A's final goods producers importing intermediates and IT services in this volume from country B.

5.3.3 Trade in IT Services

Following the IT Aadministration Rights approach of Sect. 3.4, the organizational relationship between IT service companies and intermediate suppliers follows the organizational structure between intermediate and final goods producers. Therefore, in the capital-intensive sector Y, integrated IT service departments are employed, whereas in the labour-intensive sector Z, independent IT service companies prevail.

Regarding the intermediates, it was assumed that all variants are consumed in both countries according to their share of worldwide expenditures. Accordingly, the IT services departments and IT service providers have to build up IT systems to organize the supply chain between intermediates producers and final goods producers or the integrated producers in both countries. The distribution of IT services employed domestically and abroad follows the distribution of variants produced.

The total factor demand for IT services is also incorporated in the lines of the FPE parallelogram in Fig. 5.3. Please note that trade between country A and B is balanced regarding the sum of intermediate trade and IT services, but trade in services has not been balanced. More specifically, the pattern of trade in IT services in shown in Fig. 5.4.

The x-axis represents the total number of variants produced in the world economy and also shows the number of variants produced in Y and Z industries worldwide. The diagonal represents all factors demanded by capital and labour used for IT services and is 'a fraction' of the diagonal presented in Fig. 5.3. In the left hand part, the lines $|O^A Y|$ and $|YO^B|$ from Fig. 5.3 are copied. In Fig. 5.4, they have no direct economic interpretation, but are used to determine the relative allocation of the IT services.

Using graphical analysis again, the dots on the diagonal represent the shares of total IT services. Sector Z is split analogously. Figure 5.4 shows the deployment of IT services in this framework[8]:

---

[7] This interpretation changes the meaning of the lines of the parallelogramm. No variants are added anymore, but they show the total volume of the sector, and a part of the line does not represent a smaller number of variants but a share of total volume (including all different variants).

[8] The definitions for the several types of outsourcing are presented in Table 2.9.

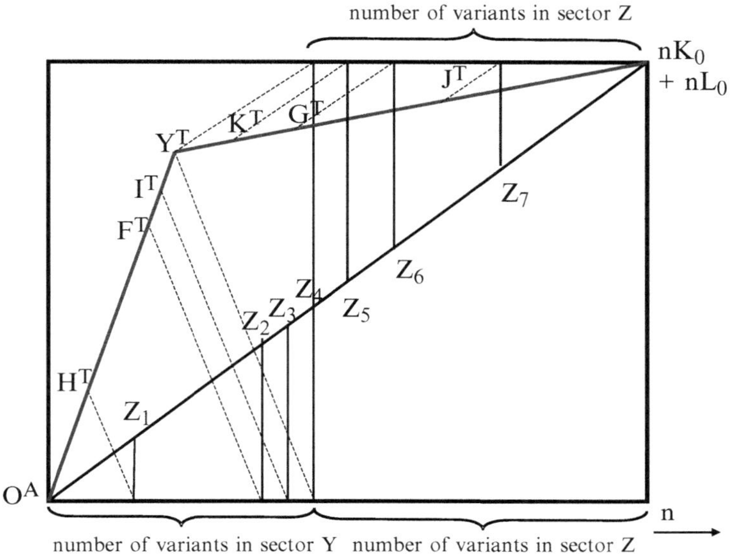

Fig. 5.4 Trade in IT services

- $|0Z_1|$: Domestic in-house production. IT services of integrated companies for domestic use only.
- $|Z_1Z_2|$: Offshoring from the perspective of country B. In-house exports of IT services from country A to country B, e.g. IT departments of integrated (multinational) companies are employed abroad.
- $|Z_2Z_3|$: Offshoring. In-house exports of IT services from country B to country A.
- $|Z_3Z_4|$: Domestic in-house production. IT departments employed domestically in country B.
- $|Z_4C_5|$: Domestic outsourcing. IT service companies employed in country A.
- $|Z_5C_6|$: International outsourcing. Export of IT services from country A to country B.
- $|Z_6C_7|$: International outsourcing. Export of IT services from country B to country A.
- $\left|Z_7O^B\right|$: Domestic outsourcing. IT service companies employed domestically in country B.

In this model, the trade of IT services does not depend on differences in factor costs (or productivity or factor endowments), but on the concept of accompanying domestic suppliers. Furthermore, it can be proved that a change in ζ representing the costs of IT services has no effect on the factor price equalization region: In this model, a lowering of ζ would lead to lower IT service costs per variant, a higher number of variants n, and a lower production volume, but without changing the wage-interest rate ratio described by (5.54). Doing this exercise for both countries and industries Y and Z shows that country A prefers to export in-house IT services to country B.

5.4 Summary and Interpretation

The presented model links trade in IT services to capital intensity. The consequences of the approach developed here are as follows:

- Imports and exports of IT services are rising simultaneously. Trading volume in IT services is largest between countries of similar size.
- It is proved that international trade in IT services occurs even if there are no differences in costs or factor intensities in the IT services sectors. Accompanying non-IT companies internationally is a key driver of the internationalization of IT services.
- Exports in-house and imports in-house behave in a reciprocal manner.

These results also describe different forces which influence the market structure in the IT sector. Since outsourcing of non-ICT intermediates is favoured in labour-intensive industries, final and intermediate goods producers keep their own IT systems (they do not integrate) and IT has to provide standardized interfaces. But in capital-intensive industries, vertically-integrated companies are preferred and IT has to support the proprietary systems, as proved with the Administration Rights approach of Sect. 3.4.

The prediction that in a capital-intensive industry all IT service companies are also integrated is overdrawn. It is more plausible that independent IT service companies are employed, but the integrated companies hold all administration rights. Therefore, internationalization of a capital-intensive industry implies that (domestic) IT service companies also have to internationalize – a pattern which has favoured the international expansion of SAP, for example. In labour-rich industries, however, local IT service companies are employed, which is favoured by open and standardized interfaces. In both countries, an insourcing business might be a valuable field of business specialization as well, but dominated in volume by the labour-rich country.

Chapter 6
Model: Entry of a Digital Goods Producer

6.1 Introduction

In this chapter, a theoretical framework of an access provider and producers of digital goods is presented. In Sect. 6.2, producers of pure digital goods compete for price-quality leadership. In Sect. 6.3, a platform provider of a digital, two-sided market is introduced. In both versions, a profit function for the digital producers in a closed economy is developed first. Then the economy is opened and the best choice of strategy for a digital goods producer located elsewhere in the world is deduced.

6.2 Quality Competition

6.2.1 Closed Economy

The framework is a general equilibrium model with two production stages and simultaneous Cournot and Bertrand competition which follows the approach of Grossman and Helpman (1991).[1] The model is described by the following characteristics:

- The revenues are influenced by the quality of the products.
- In the model, the marginal costs of producing digital goods become irrelevant.
- When the closed economy is opened, the aspatial character of the companies producing digital goods is stressed (see Sect. 2.2.3). The foreign digital goods producer is located elsewhere in the 'rest of the world'. The model describes its decision of entering a market of a small economy depending on expected profits.

First, the framework of a closed economy is developed. For a representative household, a standard CES-utility function is assumed:

[1] Chap. 4, Rising product quality.

M. Vogelsang, *Digitalization in Open Economies*, Contributions to Economics,
DOI 10.1007/978-3-7908-2392-9_6,

$$U = \left(\sum_{j=1}^{n} x_j^{\rho} \right)^{1/\rho} \tag{6.1}$$

with elasticity of substitution:

$$\sigma = \frac{1}{1-\rho} > 1.$$

There are n different industries j which are represented in the utility function by one final product x_j each. The final products are imperfect substitutes, which means that monopolistic competition is predominant in the final goods markets. The budget constraint reads:

$$E = \sum x_j p_j = w_G L_G + w_I L_I + \Pi \tag{6.2}$$

with L_G and L_I for the labour employed for distribution of the final goods and for intermediates production I respectively, w_G and w_I for real wage rate in the sectors, p for the prices of the n different final products x and profits Π summed up over all companies with:

$$\Pi = \sum_{j=1}^{n} \sum_{i=1}^{v_j} \Pi_{j,i} \tag{6.3}$$

The number of intermediates producers for each industry j is given by v_j. With regard to the standard assumptions of monopolistic competition (free-entry, Cournot-Competition), it results in a mark-up λ so that the prices for the final products x_j exceeds their marginal costs by:

$$\lambda = 1/\rho \tag{6.4}$$

Up to this point, all industries in the economy are symmetrical. Regarding the structure of the model, the symmetry helps to link total labour to industry spending.

Henceforth, industry $j = d, d \in \{1, 2, .., n\}$, which produces and delivers pure digital goods is in the focus. The access provider takes the role of a final goods producer. It is best here to think about a monopolistic access provider which manages its own portal where the digital goods are bundled and sold. Digital goods are content in this context and can be interpreted as intermediates.

The Internet access provider has to cover fixed costs G for providing connectivity, sales, and marketing costs. It is assumed that marginal costs of connectivity are zero, when the infrastructure (hardware, cables) is already installed. Fixed costs and variable costs of buying the digital goods add up to the total cost function C:

$$C_j = G w_G + v \pi_j x_j \tag{6.5}$$

with π_j for the price of the digital goods which is determined in (6.9). The input factor v results from the production function for the final products:

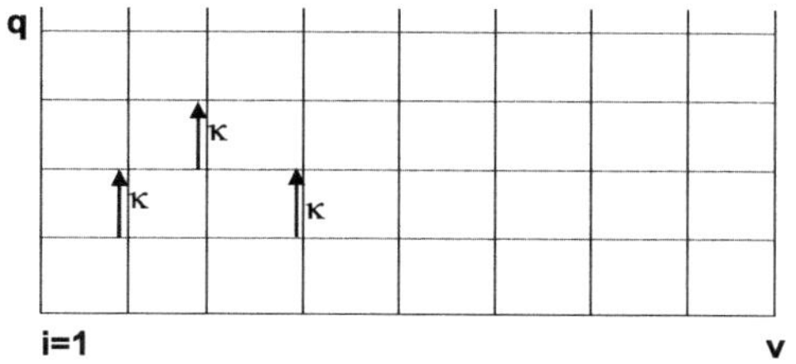

Fig. 6.1 Quality ladder for industry j
Source: Grossman and Helpman (1991)

$$x_j = \sum_{i=1}^{v_j} \frac{1}{v} z_{j,i}(q_{j,i}) \tag{6.6}$$

by assuming $v_j = v = \overline{v}$. For mathematical convenience and without loss of generality, the input factor 1/v balances the number of v different digital goods z so that for the production of x_j units, the very same number of units of the digital good are used. $q_{j,i}$ indicates the 'quality' of digital intermediate $z_{j,i}$.

The 'quality' of the digital product may be defined in terms of velocity, security, manageability, and ease of use. Following Grossman and Helpman (1991), a quality ladder as in Fig. 6.1 represents the enhancements achieved.

The horizontal axis represents the digital goods producers indexed by i. The quality level q of the digital goods is shown on the vertical axis. Improving quality means to developing a next-generation version of the digital good. By assumption, a next-generation digital good provides exactly κ times as much quality (or service) as the product of the generation before it. $\kappa > 1$ is constant and exogenous.

Fixed costs α are incurred for research and development (R&D) of a digital good $z_{j,i}$. Furthermore, the production of the digital good has a variable component for maintaining, administrative, and supporting services with a fixed input factor $1/\beta$:

$$z_{j,i} = \frac{1}{\beta} L_{j,i}^{var} \tag{6.7}$$

Bertrand competition in the market of digital goods leads to the result that only the company offering the best quality-adjusted price for a digital good i will survive. Consider two companies A and B which both offer the digital good i and produce them with quality $q_{j,i,A}$ and $q_{j,i,B}$ respectively. Quality adjusted prices Q are:

$$Q_{j,i} = \frac{\pi_{j,i}}{q_{j,i}} \tag{6.8}$$

The quality-adjusted price is the only criterion for the access provider to decide which company should supply digital good i. With equal quality-adjusted prices, it chooses the higher-quality product for convenience. Therefore, results $z_{j,i,k} = 0$ with $k \in B, C, D, ..$ when $Q_{j,i,A} < Q_{j,i,k}$ for every k. So company A sells all

$z_{j,i}$ at price $Q_{j,i,B} = \kappa Q_{j,i,A}$, which reflects Bertrand competition in this market segment.

Introducing marginal cost pricing for company B, which yields βw_I with w_I for the real wage in the digital goods-producing sector according to (6.7), and assuming that company B is second-best, the price which has to be paid by the access provider to company A yields marginal costs of B times the factor κ:

$$\pi_{j,i} = Q_{j,i,B} q_{j,i,A} = \kappa \beta w_I \tag{6.9}$$

For determining a general solution, some symmetry assumptions are introduced: (1) $\beta_{j,i} = \beta$, (2) $w_{I,j,i} = w_I$ and (3) $\kappa_{j,i} = \kappa$ for all j,i.

The amount invested in R&D is measured by α. To decide to produce $z_{j,i}$, company A has to attain a non-negative profit: $\Pi_A \geq 0$:

$$\begin{aligned} \Pi_A &= z_{j,i} \left[\pi_{j,i} - \frac{\alpha}{z_{j,i}} w_I - \beta w_I \right] = z_{j,i} w_I \left[\kappa\beta - \frac{\alpha}{z_{j,i}} - \beta \right] \\ &= z_{j,i} w_I \left[(\kappa - 1)\beta \right] - \alpha \end{aligned} \tag{6.10}$$

The equilibrium conditions for the labour market finish the description of the closed economy:

$$L = L_G + L_I \tag{6.11}$$

with:

$$L_G = nG \tag{6.12}$$

$$L_I = \sum_{j=1}^{n} \sum_{i=1}^{v_j} (z\beta + \alpha) \tag{6.13}$$

The next step is to transform (6.10) so that the profit can be expressed with exogenous variables only. Then the expected profit function of a foreign digital goods supplier can be examined when the closed economy is opened. First of all, the cost function is amended by the industry's pricing model. The expenditure per industry is calculated, which gives solutions for x and z. Assuming symmetry of all n industries, a general cost function can be determined.

With (6.5), total costs of producing and selling final products x_j are:

$$C_j = G w_G + v \pi_j x_j = G w_G + v \beta w_I \kappa x_j \tag{6.14}$$

Marginal costs account to:

$$\frac{\partial C_j}{\partial x_j} = v \beta \kappa w_I \tag{6.15}$$

In equilibrium, the price of the bundled final product (access plus content) equals the marginal costs times the mark-up factor $\lambda > 1$:

$$p_j = \lambda v \beta \kappa w_I \tag{6.16}$$

In each industry j, the difference between price and marginal costs is used to cover fixed costs:

$$Gw_G = x_j[\lambda v\beta\kappa w_I - v\beta\kappa w_I] \tag{6.17}$$

Assuming labour flexibility between all sectors $w_G = w_I = w$ leads to:

$$x_j = \frac{G}{(\lambda-1)v\beta\kappa w} = \frac{G}{(\lambda-1)v\beta\kappa} \tag{6.18}$$

The number of industries n then is given by the standard solution for demand in monopolistic competition[2]

$$x_j = \frac{p(j)^{1/(\rho-1)}}{\int_0^n p(y)^{\rho/(\rho-1)}dy}E = \frac{p(j)^{-\sigma}}{\int_0^n p(y)^{-\sigma}dy}E \tag{6.19}$$

Assuming symmetry for all i, it directly follows $x_j = z_{j,i}$ from the production function (6.6) since input coefficient 1/v balance the number of digital goods. Inserting (6.18) in (6.10) gives the profit function of company A:

$$\Pi_A = \frac{Gw(\kappa-1)}{(\lambda-1)v\kappa} - \alpha = Gw\left[\frac{1}{(\lambda-1)v} - \frac{1}{(\lambda-1)v\kappa}\right] - \alpha \tag{6.20}$$

Real wage is endogenous in the model structure and depends on labour supply and demand. When it is chosen as numeraire with $w = 1$, (6.20) expresses the expected profit Π_A in terms of exogenous variables only. The interpretation of (6.20) is intuitive:

- Technological leadership affects profits positively.

$$\frac{\partial \Pi_A}{\partial \kappa} > 0$$

- Marginal costs β are irrelevant for profits of the digital goods provider A. Mathematically, this is a direct result of the macroeconomic point of view in combination with Cournot and Bertrand competition.
- The higher the number of digital goods producers in industry j = d, the lower the profit.

$$\frac{\partial \Pi_A}{\partial v} < 0$$

[2] Symmetry with $p(j) = p$ for all $j \in \{1..n\}$ would directly lead to:

$$x_j = \frac{p(j)^{1/(\rho-1)}}{np(j)^{\rho/(\rho-1)}}E = \frac{E}{np(j)}.$$

- The higher the fixed costs of the access provider, the higher is the profit of the digital goods producer. The reason is that with higher fixed costs, the access provider has to reach higher economies of scale.

$$\frac{\partial \Pi_A}{\partial G} > 0$$

- The higher the fixed costs of the access provider, the more profitable are investments in R&D by the digital goods producer:

$$\frac{\partial(\partial \Pi_A / \partial G)}{\partial \kappa} > 0$$

6.2.2 Market Entry from Abroad

Opening the described economy changes the perspective towards a single company F outside this country. Its headquarters is located 'somewhere' in the rest of the world, which emphasizes the aspatial character of digital goods. The production function of F corresponds to (6.7); moreover, μ denotes the additional marketing and sales costs for market entry in the former closed economy, and ψ describes the probability of having success, which means becoming the quality-price leader in the Bertrand competition game. The possible strategies of market entry are[3]:

1. Strategy 1 - Trade: F enters the market and offers its own digital product to the access provider. Transport costs can be neglected with digital goods. Assuming that the product is already developed for the foreign market, which means the fixed costs α are not relevant for the decision on market entry, the profit function then yields:
$$E(\Pi_{foreign}) = \psi \left[\frac{G(\kappa - 1)}{(\lambda - 1)\nu\kappa} \right] - \mu \tag{6.21}$$
with probability ψ that the local competitor A for variety i can be squeezed out.
2. Strategy 2 – Acquisition: The foreign company F buys domestic company A. K is the price of the acquisition, ρ the discount factor for the next period's profits, and $(1-\psi)^{i_t^*}$ describes the probability that not one of the potential competitors will be successful till t.

$$E(\Pi_{foreign}) = \sum_{t=0}^{\infty} \left[(1-\psi)^{i_t^*} \frac{G(\kappa - 1)}{(\lambda - 1)\nu\kappa} - \alpha \right] \rho^{-t} - K \tag{6.22}$$

[3] For convenience, it is assumed that the company is the only potential foreign competitor and that F is 'small' in the sense that disequilibria in the balance of payments can be neglected.

3. Strategy 3: Do nothing – No market entry

$$\Pi_{foreign} \leq 0 \tag{6.23}$$

Equations (6.21) and (6.22) predict the internationalization strategies of digital goods producers, provided $\Pi_{foreign} > 0$ is expected:

- A high, G and a low λ are preferred by foreign digital goods producers to enter. Interpreting the necessary marketing and sales costs G as a proxy for the size of the country means that larger countries are more attractive for foreign digital goods producers.
- Strategy 2 (acquisition) is preferred to strategy 1 (trade), when the technological leadership of the domestic price-quality leader is dominant and seems sustainable, so that the risk for the foreign company of not achieving the leadership position itself is high (low ψ). Or, expressed in terms of the 'theory of internalization' (see Sect. 3.3.1: higher risks are adjusted by choosing a higher entry mode.

6.3 Two-Sided Markets

The Bertrand quality competition in the previous section neglects the importance of the customer base in digital goods markets. The theory of two-sided markets which introduced in Sect. 3.2 can be used to take into consideration the utility of buyers, which is related to the number of sellers and vice versa more explicitly. As in the preceding section:

CES utility function:

$$U = \left(\sum_{j=1}^{n} x_j^{\rho} \right)^{1/\rho}$$

budget constraint:

$$E = \sum x_j p_j = w_G L_G + w_I L_I + \Pi; \ \ \Pi = \sum_{j=1}^{n} \sum_{i=1}^{v_j} \Pi_{j,i}$$

and mark-up rule:

$$\lambda = 1/\rho$$

remain valid.

Here, in industry $j = d$, the Internet access provider offers v digital platforms to its customers. Economically, the digital platforms are separate two-sided markets; the price-elasticity of substitution between them is zero.

The labour input of providing the digital platform is fixed and amounts α. Labour market conditions are given by (6.11), (6.12) and:

$$L_I = \sum_{j=1}^{n} \sum_{i=1}^{v_j} \alpha \tag{6.24}$$

I assume that company A has already reached the entry-impeding levels of b_m and s_m as described in Sect. 3.2.3. According to (3.1), the number of interactions amounts:

$$I = (H_b(r_b) + H_s(r_s))s_m b_m \tag{6.25}$$

with r_b and r_s for the transaction fees determining the matching probabilities H_b and H_s. A seller transacts with every H_sth buyer and every H_bth seller is a counterpart for a buyer. Transaction fees are a result of the maximization process (see Chap. 3.2) and are exogenous here.

The resulting profit function of a provider of a two-sided market yields:

$$E(\Pi_A) = \Pi_A = (r_b H_b(r_b) + r_s H_s(r_s))s_m b_m - \alpha w_I \tag{6.26}$$

Taking into account love for variety on industry level, all consumers use products offered by this industry. The number of customers is approximated by the labour force $L^s = L$ in equilibrium. The number of transactions:

$$s_m b_m = \frac{L(L-1)}{v} \tag{6.27}$$

depends on $L(L-1)$, since every consumer can act as a buyer and seller on one of the v platforms. This implies single-homing.

I assume that the transaction fees are passed through by the access provider. Transaction fees are fixed costs from the perspective of the access provider. The customer base with s_m sellers and b_m buyers is locked into the two-sided market offered by company A, so that the access provider has no power to renegotiate the passing-through agreement. Behind this is the assumption that company A owns the property rights of the customer base, so that it could migrate with its platform to another access provider, e.g. from an Internet access provider to a mobile access provider in the industry context of this model. Monopolistic competition on industry level implies that revenues compensate the sum of fixed costs.

The cost function C of the access provider reads:

$$C(v) = v(r_b H_b(r_b) + r_s H_s(r_s))\frac{L(L-1)}{v} + v\alpha w_I + G w_G \tag{6.28}$$

The number of 'final goods' offered by the access provider conforms to the number of digital platforms in its portfolio:

$$x = v \tag{6.29}$$

With wage $w = w_G = w_I = 1$ as numeraire, the marginal costs of the Internet access provider yield:

$$\partial C(v)/\partial x = \alpha \tag{6.30}$$

Due to monopolistic competition, the price amounts:

$$p_i = \lambda \partial C(v)/\partial x = \lambda\alpha \tag{6.31}$$

The fixed costs of the Internet access provider G are covered by the mark-up:

$$v(r_b H_b(r_b) + r_s H_s(r_s))\frac{L(L-1)}{v} + G = x\,[\lambda\alpha - \alpha] = x(\lambda - 1)\alpha$$

Transforming:

$$v = x = \frac{(r_b H_b(r_b) + r_s H_s(r_s))L(L-1) + G}{(\lambda - 1)\alpha} \tag{6.32}$$

Inserting in (6.26) with (6.27) leads to the expected profit function of the provider of the two-sided market, company A:

$$E(\Pi_A) = \frac{(r_b H_b(r_b) + r_s H_s(r_s))L(L-1)(\lambda - 1)\alpha}{(r_b H_b(r_b) + r_s H_s(r_s))L(L-1) + G} - \alpha \tag{6.33}$$

The expected profit $E(\Pi_A)$ is non-negative, when:

$$1 \leq \frac{(r_b H_b(r_b) + r_s H_s(r_s))L(L-1)(\lambda - 1)}{(r_b H_b(r_b) + r_s H_s(r_s))L(L-1) + G} \tag{6.34}$$

A necessary, but not sufficient, condition for that is:

$$\lambda > 2$$

which also implies:

$$1 < \sigma < 2$$

A small elasticity of substitution on industry level as defined in (6.1) is necessary for a positive profit of the provider of a digital two-sided market platform. In other words: consumers must appraise the Internet industry so that they do not easily substitute Internet access by consumption goods of other industries, in order that the provider of the digital two-sided market gains non-negative profits. When $\sigma < 2$ is true, the profits rises with the size of the labour force L.

The above condition (6.32) defines the maximum number of two-sided platforms v the industry can bear. With a shrinking σ, the number of digital platforms diminish and the profit of the remaining companies rise. Customer preferences on industry level also have a significant influence on companies' profits, a result which completes the microeconomic approach of Sect. 3.2. Furthermore, the result that $\sigma < 2$

is a necessary condition for a non-negative profit does not change with the values of r_b and r_s. A small elasticity of substitution on industry level is a precondition that a locked-in customer base will be rewarded. In this sense, the sector hands over its elasticity of substitution to the firm level.

Opening the Closed Economy

Following the approach of Sect. 6.2.2, a foreign provider F of a two-sided market with headquarters elsewhere in the world evaluates risks and opportunities of a market entry in A's home market:

- Strategy 1 – Trade: Assume that F does not get access to the local customer bases. Then F would have to build up its own local customer base, which will not be possible when the entry-deterring levels b_m and s_m have already been reached. An exception might only be possible when the foreign customer base b_m^f and s_m^f is so attractive for domestic customers that they can increase their utility by changing the platform.
- Strategy 2 – Acquisition: Choosing an acquisition strategy, the foreign company F can match its own customer base b_m^f and s_m^f with the local customer base b_m and b_s. Profits rise due to a rising number of international interactions with F's customers to:

  $$I = (H_b(r_b) + H_s(r_s))s_m b_m + H_b(r_b)(s_m b_m^f) + H_s(r_s)(b_m s_m^f)$$

  This leads to two conclusions:

  1. **Higher M&A volumes:** Discounting future profits, the foreign competitor may pay a higher price for the acquisition of A than the domestic customer base might be worth on its own. For recent examples of mergers and acquisitions see Sect. 3.2.1.
  2. **Higher variety in digital markets, but overall number of industries decreases:** Assume that all domestic digital platform providers are acquired by an international competitor and (a part of) the additional transaction fees are cleared by the domestic access provider. Then, in this model, the nominator of (6.32) increases, which raises the number of digital platform providers v. On the other hand, the number of different industries n (or final products) decreases. With two-sided markets, international competitors improve the number of different digital platforms in the new equilibrium, but the overall variety of different industries may go down, i.e. the country specializes in fewer sectors. Industry competition intensifies by market entry of a foreign competitor driven by preferences of domestic consumers. This result reflects the characteristic of two-sided markets, that the utility of the buyers is caused by the number of sellers and vice versa.

Summarizing, in two-sided markets, foreign companies will enter by acquisition when they have missed building up their own customer base from the very beginning.

6.4 Interpretation

The integration of digital goods in a theoretical framework shows that economic interdependencies rely not only on the general characteristics of digital goods, but also on the concrete modelling of the relationships between the digital goods producers, the access provider, and the rest of the economy. Although both approaches presented in Sects. 6.2 and 6.3 emphasize the importance of being in a leadership position in digital markets, some crucial economic parameters have quite opposite impacts. The impacts on the profit functions of the digital goods producers are shown in Table 6.1.

These results show the importance of taking into account general industry or macroeconomic variables when evaluating the impact of digitalization. In the first approach (Sect. 6.2), quality competition between pure digital goods $x_j = z_{j,i}$ is assumed: the number of the final good (Internet access plus content) equals the number of digital goods produced by each of the v digital goods producers. The (potential) producers of digital goods compete in a price-quality game, and digital goods are priced according to their marginal costs plus a margin for technological leadership. Figuratively, the scale effects of producing the pure digital goods are passed through on industry level. Higher fixed costs G and a higher elasticity of substitution augment the supplies of x_j and $z_{j,i}$.

However, in the following Sect. 6.3, with a two-sided digital platform, the access provider could not gain from competition between the digital platform providers since it is assumed that the customer base of b_m and s_m is bound to A. The industry supply (access plus content) equals the number of platforms $x = v$, but not the number of transactions. The sum of transaction fees are fixed costs from the perspective of the access provider. As a result, only the digital platform provider benefits from the scale effects of the digital platform, a lower σ and a lower G expatiates profits.

Internationally, the two-sided platform approach gives a strong argument for mergers and acquisitions of digital platform providers: the merged customer base is value-enhancing, the value of a company supplying only domestically may rise when the economy (the market) is opened. For the pure digital goods producer, the choice between trade or acquisition is a matter of risk: higher risks are adjusted by

Table 6.1 Digital goods in the general equilibrium framework

Positive impact on profit of A	Sect. 6.2:Pure digital goods	Sect. 6.3:2-Sided digital platform
Elasticity of substitution	Higher σ	Lower σ
Fixed costs of access provider	Higher G	Lower G

choosing higher entry modes. Both approaches emphasize the strategic character of FDI in digital markets, and that large countries (approximated by G in Sect. 6.2 and measured by L in Sect. 6.3) are preferred to small countries to enter. With two-sided digital markets, entry of a foreign competitor may lead to a utility-driven restructuring process of the domestic economy.

Furthermore this could be a new driver of international economic divergence, when the expected size of the digital home market yields considerable first-mover advantages for innovative digital firms. However, in turn this could be an impulse for small economies to consider more urgently the options of regional economic integration.

6.4.1 Potential Expansion

Despite these insights, a number of questions remain open or are not targeted by the presented models. It may be criticized that opening a closed economy seems to be somewhat artificial, although it is a standard technique in trade theory. Furthermore the model gets rid of its general equilibrium characteristic (see also Sect. 7.2). In the closed economy the markets are balanced, domestic profits effect domestic demand. In the global case, the number of industries has to adjust according to the balance of payment including the country-specific distribution of profits, but this relationship is not in the focus here. On the other hand, it is a matter of fact that digital products are often supplied in a local/limited market first (e.g. defined in terms of language) before the producer attacks the global market and profits are scaled up. Also, the models do not take into account that the law of gravity (e.g. because of local preferences) may be true for digital services also. But a study of Blum and Goldfarb (2005) shows that this is a specific phenomenon bound to the type of goods/services offered. Furthermore, the model framework could be criticized as being rudimentary in details, e.g. the competition behaviour between the digital goods producers in an industry is simplified.

But in return for these shortcomings, the interdependencies between general macroeconomic parameters and the specific sector characteristics are highlighted.

Chapter 7
Comparison of the Models

7.1 Framework

Digitalization is visible in networks, IT services, and digital goods. Their special characteristics are introduced in Sect. 2.2, and theoretical foundations in Chap. 3. In this chapter, the presented general equilibrium models of Chap. 4 (networks), Chap. 5 (IT services), and Chap. 6 (pure digital goods and two-sided digital platforms) are compared.

The aim of the analysis of digitalization in the theoretical frameworks is to extract interdependencies between digitalization and the structures of an open economy. The approach of this thesis differs from other work about digitalization, which review specific digital markets in a partial equilibrium. Supply and demand are embedded in a general framework, where principle structures (e.g. factor endowments) determine prices and quantities.

The monetary side of an economy (e.g. money supply and demand) is not within the scope of the models. Labour markets are described rudimentarily as permanent labour market equilibrium is assumed. Here, the presented models follow the neo-classical paradigm of general equilibrium models. The use of neo-classical assumptions allows one to focus on the good characteristics of networks, IT services, and digital goods, without having to take care of repercussions on labour market structures. Of course, this masks the effect of faster human capital accumulation through digital goods, but from the theoretical point of view, the human capital effect of digital goods is not expected to be different from standard human capital accumulation theory.

The models presented in Chaps. 4–6 focus on the real side of an economy and highlight the relationships between final goods (or industries) and intermediate producers. This structure is chosen because final and intermediates producers can easily be reinterpreted, e.g. as access and digital goods producers.

In the four variants of the framework, the assumption of a representative household with CES preferences who loves variety is used. Alternatively, the interpretation of a CES production function which determines the quality of a final good is used in Chap. 4. This paves the way for the monopolistic competition tool. With this standard assumption, it becomes possible to break down aggregate income on

M. Vogelsang, *Digitalization in Open Economies*, Contributions to Economics, 141
DOI 10.1007/978-3-7908-2392-9_7,

industries. In this sense, monopolistic competition is the connecting hinge between the macroeconomic view of the economy and the microeconomic magnifying glass on the topic of interest.

7.2 Differences in Theoretical Structures

Following the concept of a framework with different designed models, the distinctive, economic features of networks, IT services or digital goods can be worked out.

Because IT services are heterogeneous, the question of the structure of trade in IT services is in the foreground. To explain the structure of trade, a 'structural input' is needed; in Chap. 5, this is the labour-capital endowment ratio, best to be explored in a two country environment.

Networks and digital goods (at least in a considered sub-segment) are homogeneous and other questions are emphasized. Here, no 'structural input' has to be considered, so the models of Chaps. 4 and 6 only rely on the input factor of labour.

The models of Chaps. 4 and 5 are General Equilibrium (GE) models so that a unique equilibrium is existent which balances the forces, determines prices and depends on tastes, technology, and the distribution of wealth basically.[1] The desired IT administration right can be interpreted as one of these 'primitives'. Furthermore, so that a general equilibrium in the sense of Arrow-Debreu exists with increasing returns, some 'survival assumptions' have to be imposed, for example that the marginal returns, though they are increasing, are bounded or inputs and outputs are distinguished a posteriori (Bonnisseau and Jamin (2004)). In Chap. 4

Table 7.1 In the focus of GE models

	Networks (Chap. 4)	IT services (Chap. 5)	Digital goods (Sects. 6.2 and 6.3)
Aim to explain	Spread of international networks and the international trade of goods	Structure of Trade in IT Services: offshoring vs. International Outsourcing	Market Entry Entry modes: trade or acquisition / FDI
Goods characteristics	Homogeneous	Heterogeneous	Homogeneous in a specific sub-market, heterogeneous in industry.
Focusing on	Global Economy	Two countries	1. Closed economy 2. Opening for RoW
GE	Yes	Yes	In the closed economy case

[1] For an overview about the development of the General Equilibrium theories led by Walras, Arrow and Debreu, see Darrell and Sonnenschein (1989) for example.

Table 7.2 Linkages between macroeconomic and sector-specific variables

Networks (Chap. 4)	IT services (Chap. 5)	Pure Digital Goods (Sect. 6.2)	2sided digital platforms (Sect. 6.3)
Intermediate producers supply domestically or internationally. Internat. trade needs communication and coordination.	IT administrations rights (incomplete contracts + accompanying companies abroad)	Industry-level parameters influence profit expectations.	Industry-level parameters and implicit utility of number of transactions influence profit expectations.

this 'survival assumption' is the monopolistic competition. In Chap. 6.2 the General Equilibrium exists in the closed economy. Although digital goods are considered, a bound is introduced via the assumptions of quality-based pricing and the introduction of an access provider as an intermediate producer (or reseller) which is restricted by monopolistic competition. Without these limits it would not be possible to determine equilibrium prices. This is also the case after the economy had been opened to foreign competition. It is assumed implicitly, that the opening does not change the basic industry structure of the economy under consideration.

After the choice of the framework, utility and production functions, input factors, and industry structure, the most important feature is the link between the macroeconomic variables and the network, IT services and digital goods providers. Table 7.2 shows the instruments used.

Following the neo-classical tradition of GE models in the network model of Chap. 4, the zero-profit condition (interpreting personal income of the entrepreneur as labour income) is used to solve the trade-off between communications and coordination costs on the one hand, and scale effects of production on the other. In the model about IT services, expected profits play a particular role to determine the Nash equilibrium with incomplete contracts. In both models of Chaps. 4 and 5, further assumptions would be necessary to describe the profitability of the network and IT service providers. In contrast to these, the two approaches about digital goods and platforms in Chap. 6 refer to the arguments that digital market characteristics allow attainment of a leadership position which can be used to earn monopolistic rents.

7.3 Main Conclusions

The main conclusions refer to the interplay between digitalization and internationalization of an economy. Table 7.3 shows the drivers of internationalization of networks, IT services, and digital goods providers.

Table 7.3 Drivers of internationalization

	Networks (Chap. 4)	IT services (Chap. 5)	Digital goods (Chap. 6)
Key drivers for internationalization	Scale effects in production	Accompanying multinational companies or supporting outsourcing solutions	Searching for new markets
Incorporated in model by	Fixed costs in intermediate production	Incomplete contracts	Quality-price competition or merger of customer base
Main criterion for company's internationalization strategy	Trade-off between economies of scale and communication + coordination costs caused by internat. trade	IT systems architecture required due to endogenously-determined organizational structure	Expected profits

The main distinction between the models refers to the relationship of the providers to their customers: does internationalization mean accompanying an existing (domestic) customer when entering foreign markets, or does it mean searching for new customers? Evidence for both scenarios of Chaps. 5 and 6.2 are exemplarily given by considering two software producers.[2] SAP produces Enterprise Planning Systems and is an IT service company which accompanies its companies worldwide. Service fees (maintaining) contribute to SAP's total revenues more than subscription and licensing fees. In contrast, F-Secure produces anti-virus software and is a digital goods producer which systematically enters foreign markets to acquire new customers. Subscription and licensing fees contribute to F-Secure's total revenues more than the service fees. More case study examples related to internationalization strategies of ICT companies can be found in Dachs et al. (2007).

Now the main aspect of the theoretical models can be worked out: the effects of digitalization. Digitalization has three effects on the models presented.

- Digital goods with special characteristics as described in Sect. 2.2.3 enter the scene.
- Decline of costs in providing network and IT services.
- New methods and possibilities of organizing work and information flows inside and between companies as described with the IT administration rights of Sect. 3.4 in condensed form.

[2] For the discussion about the classification of software as a good or a service, see Sect. 2.2.4.

Table 7.4 Effects of digitalization

Networks (Chap. 4)	IT services (Chap. 5)	Digital goods (Chap. 6)
Digitalization lowers communication costs. Trade volume increases and the number of companies which offer their products only locally diminishes. The globalization of the value chain favours investment in communications networks which will lower communications costs further.	Digitalization allows distribution of IT administration rights internationally, leading to outsourcing and offshoring (= internat. trade) of IT services. IT administration rights are an important component of property rights. Companies can implement international organizational structures more easily and stringently. A lowering of the IT service costs leads to a higher number of companies in capital- and labour-intensive industries.	With digital two-sided platforms, an increase of international mergers and acquisitions can be expected. Large markets are first exposed to foreign competition. Furthermore, acquisitions may lead to a higher, utility-driven degree of specialization of the country.

The consequences of these manifestations of digitalization are reflected by the structure of the models directly, or can be deduced by the use of comparative-static techniques. The results are summarized in Table 7.4.

Although digitalization appears with different faces, the three models have in common that digitalization allows and favours the split of the value chain, also with traditional goods. Digitalization is a driver of internationalization and vice versa, as the General Equilibrium models prove. The consequences of digitalization for a specific country may differ according to factor endowments and country size, technology, and preferences, which directly or indirectly determine the market conditions of networks, IT services, or digital goods. The direction of trade specialization, innovation systems, labour market conditions, human capital accumulation, etc. may also play a role. The effects of digitalization essentially depend on the fundamental structures of an economy.

The General Equilibrium models of Chaps. 4–6 show that networks, IT services, and digital goods are interweaved with the rest of the economy and are influential for international relationships. Digitally induced globalization is at work, which has consequences for the design of economic policy.

The topics mentioned in Table 7.5 span a set of economic policy options which rely on digitalization. Regulation and competition policies play an important role. Regarding networks the interconnection regime is crucial, most recently questions of New Generation Networks (NGN) and Network Neutrality are discussed (see Sect. 9.1). Regarding IT services the questions of standards is important, which might also be a part of a (national) strategy to support IT service companies (see Sect. 9.2). The innovation argument emphasized with digital goods refer to a broad range of policy topics, above them intellectual property rights are crucial for the diffusion of digital goods and companies (see Sect. 9.3).

Table 7.5 Lessons for economic policy

Networks (Chap. 4)	IT services (Chap. 5)	Digital goods (Chap. 6)
The spread of networks favours international trade. Questions of interconnection matter also from the macroeconomic point of view.	The internationalization of IT service companies can also be promoted by the internationalization of non-ICT companies. Furthermore, there is little reason to fear that IT services of capital-intensive (European) industries are outsourced to third countries extensively.	Innovation systems are crucial. Digital companies have to occupy leadership positions, either in price-quality competition, or in terms of a customer base bound.

The policy instrument associated with networks, IT services and digital goods influence the competitive power of single companies and countries as well. National policy perspectives might be in tension with a global policy approach, which is on the track with several international international institutions and contracts. The WTO agreements (GATS, GATT, TRIPS, etc.) are important (see Sects. 8.2, 9.2.3, 9.3.3).

Finally, convergence between telephone, internet and television networks change the traditional assignment of policy instruments. Therefore the new issue arise to develop a consistent regulatory approach which is flexible enough to take into account the macroeconomic effects of digital policy, so that the approach is supported internationally. To pave the way, Part III focuses on policy aspects which are related to networks, IT services, and digital goods.

Part III
Economic Policy

Chapter 8
General Political Perspectives

In Part II of this study, several theoretical approaches of internationalization strategies of network providers, IT service companies, and digital goods producers have been presented. Some clues for economic policy have been summarized in Sect. 7.3. Now, in Part III, a closer look at economic policy as a driver of the international diffusion of networks, IT services, and digital goods is presented.

8.1 Economic Policy: Principles, Objectives, and Instruments

8.1.1 General Goals

Economic policy should always take into consideration the meta-objectives which determine the overall welfare of a society. According to Welfens (2007d), these are peace, freedom, wealth, and fairness. Wealth can be further defined as income per capita, availability of public goods, leisure time, and natural resources.

The policy instruments discussed in this chapter can be assigned to the meta-objectives mentioned, as Table 8.1 shows. Political agenda-setting is to some extent a dynamic process as society changes. The most recent change is described by the term "information society". An information society can be characterised by its environment, readiness and usage of networks, IT services and digital goods, for example (see also Sect. 2.1.3 and Table 2.5).

8.1.2 Principles of Economic Policy in the Digital Age

In the context of networks, IT services and digital goods, three principles play an important role to build up a consistent policy framework:

- Layer-consistent governance
- Technology neutrality of regulation
- Non-discrimination/network neutrality.

M. Vogelsang, *Digitalization in Open Economies*, Contributions to Economics, DOI 10.1007/978-3-7908-2392-9_8, © Springer-Verlag Berlin Heidelberg 2010

Table 8.1 Economic policy and meta-objectives

Topic	Section	Primary meta-objective
Overcoming the digital divide	8.1.3	Fairness, wealth
Media, plurality and culture	8.1.4	Freedom
Data privacy and safety	8.1.5	Peace (national safety), freedom
Regulation and competition policy	8.1.6	Wealth, fairness (equal opportunity), freedom (freedom of competition)
Innovation, economic growth and employment	8.1.7	Wealth
Trade policy	8.1.8	Wealth, freedom (freedom of trade)

The principle of layer-consistent Internet governance means to minimize layer crossings in regulatory practice (UNCTAD (2006)). The layers are shown in Table 2.1. The principle requires that an economic problem existing in a specific layer (e.g. the last-mile problem in the physical network layer; see Sect. 9.1) is tackled by a policy instrument which only affects this layer.

Whereas the layer-consistent governance is strongly related to a technological perspective, the principle of technology neutrality rather employs a bird's eye view. OECD (2006f) states that technology neutrality aims to "provide coherent regulation for all transmission networks and services including broadcasting, regardless of the underlying technology by abolishing vertically integrated regulation that links regulation to underlying network technologies".

The principle of non-discrimination advocates the rights of users and buyers of services and digital service. For example, Michael Powell, Chairmen of the Federal Communications Commission (FCC), proposed a set of non-discrimination principles in February 2004. According to UNCTAD (2006), Internet users should have the:

- Freedom to access content
- Freedom to run applications
- Freedom to attach devices
- Freedom to obtain service plan information.

When applied to networks, the non-discrimination principles lead to the requirement of network neutrality, which is picked up subsequently.

Besides these three principles which can be applied especially to evaluate instruments for networks, IT services and digital goods markets, the whole set of political evaluation criteria is still valid with digitalization, e.g. the design of economic policy should be rational, consistent, and pay attention to the principle of subsidiarity (Welfens (2007d)). Also, the efficiency criteria, which are derived from welfare economics, can be used to assess regulatory and competition policy, for example (Dewenter and Haucap (2007)). They are as follows:

- Productive efficiency: Companies adopt the least-costs method of production.
- Allocative efficiency: Resources are employed on goods and services so that the maximum benefit to society is provided.

- Dynamic efficiency: Companies invest and innovate at an optimal rate and point in time.

The international diffusion of digital technology has not been mentioned as a goal of its own. But internationalization is a direct consequence of the principles presented above: the freedom to access content also embraces access to international content. The freedom to attach devices means to standardize software and IT services internationally. The principle of network neutrality means that every country can join the international network (the "Internet") without discrimination. Furthermore, companies are driven to operate internationally when they want to exploit economies of scale to reach productive and dynamic efficiency.

8.1.3 Overcoming the Digital Divide

The term "digital divide" means the existing discrepancies in the possibility of users access to digital communication networks, e.g. the Internet.[1] Section 2.2.4 presents some facts about the digital divide between countries. Figure 2.7 illustrates that the distribution of Internet users between countries has become more uniform in recent years. Figure 2.8 shows the strong relationship between GDP per capita and Internet use. Figure 2.9 makes clear that the digital divide is narrowing considering the rising age of technology.

Usually, it is assumed, a priori, that it is a political goal to narrow the digital divide. But why is it not argued that digital equality between countries is not necessary for a Ricardian specialization pattern of trade? On the one hand, the higher the penetration rates in all countries,[2] the higher are the resulting network effects (see also Sect. 3.1.1). On the other hand, international (digital) access to communications is a prerequisite to split up the value chain (as the fragmentation model in Chap. 4 shows), and to exploit Ricardian comparative advantages.

Economic policy to overcome the digital divide can be classified in measures to:

- Improve access, e.g. to establish IXPs (see below) or to build up a mobile access network in Africa,[3] to improve the availability of computers or to impede a universal service obligation
- Improve other individual or regional factors, such as educational skills.

---

[1] As ITU states in ITU and UNCTAD (2007), "discrepancies in access to ICTs between countries are not only measured in terms of basic penetration and access, but are taking on new dimensions in speed, mobility and capacity of access".

[2] As the number of Internet users in leading countries is converging towards the 100% limit (Iceland 74%, Sweden 76%, Iceland 87% of total population in 2005; source: World Development Indicators Database), overcoming the digital divide in this context implies offering Internet access to the entire population of all countries.

[3] According ITU (2006a), the Gini coefficient for the opportunity variable "mobile coverage" is 0.18 worldwide. Therefore, mobile networks might be a good starting point to narrow the digital divide.

Access is the prerequisite to benefit from existing digital networks, the individual and regional factors influence the frequency and manner of use. Hargittai (2003) claims that differences in the way of using the Internet should be more pronounced when defining the digital divide. Schleife (2006) emphasises that a different composition of the population causes a regional digital divide in Germany. So, "overcoming the digital divide" implies ameliorating regional and/or individual deficits. In the literature about the digital divide, which, in most cases does not consider the digital divide between countries, but the digital divide within societies, the following factors are discussed to be positively (+)/(negatively [−]) correlated with more intense use of the Internet: children in a household (+); education (+); income (+); availability of local sites (+); availability of computers (+); telephone density (+); single-person households (+); urbanity of a region (+); males (+); age (−); part-time workers (+); social support networks (e.g. others who are experienced with the Internet +) (sources, with further references: Schleife (2006), Hargittai (2003), and Hoffman and Novak (2000)). Chinn and Fairlie (2006) find out that the factors determining Internet penetration do not differ between developed and developing countries significantly. But Wellenius and Townsend (2005) point out that nearly 70% of all websites are in English and software is mostly US-centric in design and function. They claim that there is a lack of cultural media, e.g. emphasizing the languages and experiences of developing countries.

Thinking about policy instruments to overcome the digital divide within societies, the trade-off between welfare-enhancing product differentiation and welfare-enhancing network effects has to be taken into account. For example, Devescovi (2003) expects that the broadcast market will become more separated: "prodotti di massa" will be a general TV programme sent in average technology, and "i prodotti di elite" programmes, which focus on specific items and offer paid content. Tambini (2006) argues that we would live in a totalitarian state if all citizens accessed the same content. But, if some groups were extremely well-informed and communicate in a timely and strategic way, and others simply did not, the fragmentation of audience would be a greater danger to democracy than totalitarianism in the current context. Therefore, Tambini (2006) concludes that the regulator should develop instruments to protect "citizens" interests' also.

To overcome the digital divide between countries, supranational initiatives such as WSIS, a UN summit (see also Sect. 8.2.3), and national policy instruments such as infrastructure investments, prudent regulation and competition law are pivotal. In the World Summit on the Information Society (WSIS) the countries committed to countering the phenomenon of the digital divide, "by all means, to enhance connectivity and universal, ubiquitous, equitable, non-discriminatory and affordable access to, and use of ICTs" (Ponder (2007)). An example of an economically justifiable government investment in infrastructure is the establishment of a national IXP (Internet Exchange Provider; see also Fig. 9.1) in African countries. OECD (2006d) reports that data traffic from South Africa to Mauritius is routed via Pretoria-

Gauteng, New York, Paris, Salzburg, and then via satellite with a high latency time to the final destination.[4] But whatever instrument is chosen to overcome the digital divide, either between or within countries, it supports the international diffusion of networks, IT services, and digital goods.

8.1.4 Media Policy

8.1.4.1 Trade-Off Between Culture and Business

The sciences of philosophy, politics, communications and economics converge when the role, impacts, motives and independence of the media are discussed (Ludwig (1998)). Therefore, the economics of the media tend to engage in different directions: Loeffelholz (2003) mentions society, evolution of the media, networks and systems. Whereas Walter and Quiring (2006) point out the similarities, e.g. when the media are considered as a process, the formula of the communication process based on Harold Lasswell, "Who", "Says what", "In which channel", "To whom", "With what effect" corresponds to the value chain developed by Michael Porter: "Production", "Bundling", "Distribution", "Reception", "Margin".

Although it is common sense that the media have an important role to play in education, culture, society and democracy (plurality of opinions and "fourth power" to control politics and administrations), and that the media should act freely and independently, existing media systems are criticized strongly. For example: Chomsky and Herman (1988) argue that the media defended privileged groups; Adorno and Horkheimer (1969) criticize very much the quality of content.[5] Obviously, there is a trade-off between considering the media as an economic or a cultural good.

Countries have traditionally divergent views on the relationship between cultural and economic goals, as can best be seen in the trade negotiations about audiovisual services. Canada and the EU argue that AV services, because of the cultural aspects, should not be subject to liberalization rules, and, thus, have abstained from making offers in the GATS negotiation process. Others, led by the US, argue that trade in AV services promotes cultural diversity (Roy et al. (2006)).

---

[4] Related is another barrier: because of the monopolistic structure in the African backbone market, the incumbents are "free to extract monopoly rents" leading to high mark-ups on international traffic prices (OECD (2006d)).

[5] "Cinema and broadcast do not need to pretend any longer to be art. They use the truth that they are not anything else than business as ideology, which is to legitimize the trash, which they manufacture deliberately. They call themselves industries, and the published income numbers of their general managers strike down the doubt about the social necessity for their final products", wrote Adorno and Horkheimer (1969) 30 years ago. Translation by author.

8.1.4.2 Broadcast Regulation

Media observers keep an eye an broadcast programmes especially, because of its "one-to-many" technology and the power of pictures, which influences people more than written text. Governments started early to regulate the spectrum allocation, the carriage of the signals and the content of broadcasters and founded public service broadcasters such as the BBC. Today, the digitalization of the TV signal allows more efficient use of the frequency bands. For example, one terrestrial analogue channel can be replaced by four to six digital channels. This is one reason to rule out digital terrestrial transmission technologies such as DVB-T in Germany (Monopolkommission (2006) and Welfens et al. (2007)). Looking at regulatory measures, all regulators state that it was their goal to maintain different kinds of plurality: of ownership, of content diversity, and of "voices". The most important regulatory instruments are (OECD (2004a), and OECD (2007h)):

- Network-related:
  - *Must carry rules:* Cable and satellite providers are forced to transport certain programmes, e.g. regional or public programmes which are financed by taxes or quasi-taxes.[6]
  - *Must offer rules:* Channels have to offer their content to platforms if they ask for it. Only three OECD countries (Belgium, France and the UK) have imposed "must offer" obligations on their public broadcasters (OECD (2007h)).
- Content-related:
  - Advertising quota and separation of programmes and advertising.
  - Content quotas, e.g. for national productions, languages or specific content (such as religious or cultural programmes).
  - Consumer protection ("Protection of decency rules"). Two OECD countries apply the V-chip technology.[7]
- Ownership-related: General competition law or more specific media law to prevent economic market power which could lead to a bias of opinions.

Historically, national media policy has also attempted to prevent an "uncontrolled" diffusion of international content. But technological developments limit national influence, e.g. the distribution of TV programmes via satellite ignores

[6] In the European Community, "must carry" is allowed by Art. 31 of the Universal Service Directive: "Member States may impose reasonable 'must carry' obligations, for the transmission of specified radio and television broadcast channels and services, on undertakings under their jurisdiction providing electronic communications networks used for the distribution of radio or television broadcasts to the public where a significant number of end-users of such networks use them as their principal means to receive radio and television broadcasts" (EC (2002f)).

[7] The V-chip in TV receivers blocks programmes based on their rating category to protect children viewing inappropriate programmes. In the US, all TV units have been required to have the V-chip technology since January 2000.

national borders totally, as does delivery of the TV signal over the Internet which leads to a clash of the different regulatory approaches of broadcast, telecommunication and Internet (see also Sect. 9.4.3 and Tables 9.5 and 9.6).

8.1.5 Data Privacy and Safety

Shapiro and Varian (1998) emphasise that the holder of information gains economic and political power. From this point of view, the topic of data privacy has economic as well as juridical, political, and technological aspects. The protection of data is a matter in the offline world, but an even more serious topic when databases, PCs, mobile phones, etc. are connected to each other online. Three main aspects are considered: the legal collection of data (data mining), (industrial) espionage, national security, and governmental access.

The latter stands in a trade-off to human rights. Article 8 of the European Convention of Human Rights[8] (ECHR (1950)) shows that the basic conflict is not new:

- Article 8, sentence 1 states: "Everyone has the right to respect for his private and family life, his home and his correspondence".[9]
- Article 8, sentence 2 limits this right: "There shall be no interference by a public authority with the exercise of this right except such as is in accordance with the law and is necessary in a democratic society in the interests of national security, public safety or the economic well-being of the country, for the prevention of disorder or crime, for the protection of health or morals, or for the protection of the rights and freedoms of others".

In recent years, the trade-off between data privacy and interests of national security has become very apparent. Referring to the "fight against terrorism", national security agencies have extended their access on communication data such as email, telecom connections, etc.[10] %Hier könnte ich neue politische Ökonomie erwähnen.. oder nicht ? The access of the national governments to data is one aspect of this topic, the protection of data networks as an important element of national infrastructure another. For example, this is the objective of the US "National Strategy to Secure Cyberspace" which is part of the "Homeland Security" measures put

[8] Belgium, Denmark, Germany, France, Ireland, Iceland, Italy, Luxembourg, The Netherlands, Norway, Turkey, UK, Greece and Sweden signed the convention in 1950.

[9] In Germany, the term "Informationelle Selbstbestimmung" has been introduced by the Bundesverfassungsgericht to describe that it is a basic right to decide whom to forward personal data to.

[10] For example, German telecommunications and email providers must establish monitoring possibilities for the secret service. For the technical requirements, see Bundesnetzagentur (2005). A detailed overview about further instruments and the political and technological discussion about them is given on an Internet page of Heise Verlag: http://www.heise.de/ct/hintergrund/meldung/95584.

into force after the attacks of 11/9 (Whitehouse (2003)). To make clear the threats caused by organized attackers, the strategy mentions the computer worm NIMDA which destroyed files on 86,000 computers in a few days. Pre-emptive measures have been to build up a "National Cyberspace Security Response System" and to enforce education, training, etc.

Industrial espionage or cyber attacks against companies are other aspects of this topic. To prevent these illegal attacks, governments support domestic companies to protect data or impose obligations of notifying data security breaches (Schwartz and Janger (2007)). Garcia and Horowitz (2007) worry that the underinvestment in Internet security will increase if the ratio between the social value and the revenues of Internet providers from data security grows. Following standard microeconomic argumentation, then regulation might be necessary. Furthermore, citing a MessageLabs presentation, ITU and UNCTAD (2007) mention that there is a "cybercrime ecosystem": viruses install spyware and create botnets, which are used to send spam. Spammers pay for email addresses from spyware creators.

A related aspect is the data mining carried out in accordance with law. For example, content providers such as Google collect and analyse data about usage and preferences. Lessig (1999) has foreseen in 1999: "The invisible hand of cyberspace is building an architecture that is quite the opposite of what it was at cyberspace's birth. The invisible hand, through commerce, is constructing an architecture that perfects control – an architecture that makes possible highly efficient regulation". The opposite point of view argues that the provision of user-specific content according to his preferences increases the utility of online services.

Summarizing, data security and privacy are usually taken to be national tasks, although the threats are international. So, on the one hand, the diffusion of IT services is fostered because special attacks (e.g. computer viruses, worms, etc.) are international. On the other hand, data access is a part of the IT administration rights, which have a value of their own (see Sect. 3.4). Fears that the value of the IT administration rights diminishes and/or private data could be misused constrain the international diffusion of IT services.

8.1.6 Regulation and Competition Policy

8.1.6.1 Conceptual Issues

Concepts and interpretations of technical terms of regulation and competition policy differ slightly across countries and between economists and lawyers. For example, the term "Antitrust" is used in US literature, and "Competition Policy" by European authors predominately. Second, market power is a pivotal concept as the existence of market power is a prerequisite to start investigations whether regulatory or antitrust remedies as necessary. Kapolow and Shapiro (2007) point out that the legal approach of viewing market power is dichotomous: market power as present or absent. Whereas the economic approach interprets market power as a matter of degree. The third example of divergence refers to the different schools of thought: in Europe "too high" prices for consumers can be an abuse of market power; in the US this "corpus delicti" is not existent in law (Motta and Polo (2005)). Fourth, economists do not always agree on market definitions and the relevance of theoretical concepts: Should markets be defined as narrow or broad? Or: Are markets always contestable, at least in the long run, as claimed by the so-called Chicago School of thought? And: What role should contestability[11] play for the decision about remedies. Fifth, it is usually assumed that the prospect of exploiting monopolistic rents mainly fosters R&D efforts.[12] But newer research states that the relationship between competition and innovation might be more an inverted U-relationship (Aghion et al. (2002)): when competition is very intense, the difference between pre-innovations rents and expected post-innovations rents is high. So, also, competition increases the incremental profits and gives an incentive to innovate. Finally, concepts of regulation and competition also behave dynamically, e.g. they are influenced and changed by new ideas and/or research results. For example, it is argued that a credible threat of regulation could be sufficient[13] to prevent monopolistic price setting and to sustain competitive forces.

Accepting these shortcomings, Table 8.2 gives an overview of differences between regulation and competition policy. Regulation can be interpreted as a special case of competition policy. Arino (2006) emphasises that competition law can be a substitute, can complement, or act as a trigger for regulatory developments. However, if the regulation of a sector ends (what is claimed to be a goal by regulatory authorities), it is nevertheless subject to general antitrust law. Table 8.2 shows the similarities and differences (sources: Motta and Polo (2005), Briglauer and Reichinger (2008), EC (2007c), Stumpf (2006), and Streel (2007)).

Special cases are mergers and acquisitions, because they are an ex ante application of competition law. Horizontal mergers are usually considered to restrain competition; vertical mergers can also be welfare-enhancing, e.g. when the problem of "double marginalization"[14] is avoided. If the sector is regulated, regulatory policy also plays a role. Carlton and Picker (2007) argue that regulators prefer to avoid vertical integrations between competitors (e.g. in network economics, competitors of an incumbent have an incentive to merge when interconnection prices are regulated

[11] Proponents of the contestability theory (see Knieps (2003) for example) argue that a natural monopoly plus irreversible sunk costs should be the only preconditions for regulation. Briglauer and Reichinger (2008) argue that, instead, the theory of contestable markets can "by no means provide guidance for ex ante regulation or competition policy". One of their arguments is that the contestable markets theory does not take into account price-setting reactions of the monopolists if a (potential) competitor wants to enter.

[12] For example, Baake et al. (2007) propose to impose restrictions on new markets only after 4 years from the date of emergence of that market.

[13] But according to Haucap et al. (2007), this model of a "light-handed regulation" has neglected its promise.

[14] Double marginalization arises when producer and retailer exploit their monopolistic power in a specific market. See Pepall et al. (2005), for example.

Table 8.2 Regulation and competition policy

	Competition policy	Regulation
Which markets?	All markets	Industry-specific
Deals mainly with	Unsatisfactory firms' behaviour. Structural conditions of markets are compatible with normal functioning of competition	Unsatisfactory market structures. It cannot be expected that competitive forces work without problems, e.g. a natural monopoly or a transitory phase after opening of state-owned monopoly
When?	Ex post (usually) Ex ante in the case of M&A	Ex ante (usually)
In the case of (EU definitions)	Abuse of a dominant position. Prevention, restriction or distortion of competition (remark: the creation of market power through innovations, investments, etc. is not forbidden) Merger or acquisition plans	Susceptible markets: – High and non-transitory barriers to entry – Lack of dynamic trends towards effective competition – Insufficiency of competition law
Involvement	Occasional	Long-run and continuous
Main instruments	Punishment in the case of misuse (e.g. a forbidden cartel). Ex-ante approval of merger and acquisitions	See Sect. 9.1.3, e.g. to compel incumbent to grant access (Interconnection) at regulated prices

above marginal costs). Then the new firm could become a potent political force in regulatory battles. That is "why it was important for the regulators to constrain the forces of competition" (Carlton and Picker (2007)).

Although competition and regulatory policies should be designed to protect long-term competition, there is a risk of interventions failing. The potential shortcomings are (Willig (2004)):

- High-tech markets move too rapidly and unpredictably for government intervention
- Government interventions lead to delays and uncertainties which are dangerous for high-tech companies
- Intervention measures are short-term; more important is long-term competition
- "Endogeneity problem": (regulated) prices determine quantities which also influence costs (Dewenter and Haucap (2007))
- Economies of scale and scope are extremely important for high-tech products; interventions based on static views of concentration and competition might be counterproductive.

So the speed of development and convergence of networks, IT services and digital goods implies new challenges for antitrust and regulatory agencies (see also Sects. 9.4 and 10.2). Or, as Quah (2002), p.25 puts it: The digital goods markets "imply ever-escalating tensions between ex post efficiency and ex ante incentives".

8.1.6.2 Modus Operandi in EU

As mentioned above, the economic concepts used differ slightly across countries. To describe one framework precisely, I focus briefly on EU's methodology in this section. A prerequisite for the application of remedies according to both, competition law and regulation, is the proof of significant market power (SMP).[15] Competition policy and regulation converge in Europe (Streel (2007)). However, a major difference remains: SMP analysis in the case of competition law is ex post, e.g. looks on previous behaviour of companies, whereas SMP within regulation has to be persistent ex ante (EC (2002a)).

The Framework Directive for electronic communications networks and services (EC (2002e)) defines SMP in Art. 14(2): "An undertaking shall be deemed to have significant market power if, either individually or jointly with others, it enjoys a position equivalent to dominance, that is to say a position of economic strength affording it the power to behave to an appreciable extent independently of competitors, customers and ultimately consumers." SMP and dominance are used equivalently (also see the fourth line in Table 8.2). The SMP Guideline (EC (2002a)) specifies that demand-side substitution and supply-side substitution should be taken into account. Potential competition is regarded as a third source of "competitive constraint on an operator's behaviour".

In antitrust estimates, supply- and demand-side substitution elasticities are also used to identify relevant markets.[16] In the case of regulatory policy, relevant markets which are susceptible to ex ante regulation are identified initially by the European Commission. Three criteria must be cumulatively satisfied:

1. High and non-transitory barriers to entry, e.g. an incumbent dominates a market which is characterised by economies of scale and high sunk costs.
2. Lack of dynamic trends towards effective competition. Briglauer (2007) argues that the adjective "effective is used to indicate a reality-based deviation from the neo-classical ideal of perfect competition". He claims that "effective competition" is not defined precisely, and argues that the "Market Structure – Conduct –

[15] An exception is the interconnection obligation in the EU's Access Directive: The imposition of an interconnection obligation does not depend on a finding of SMP in a particular market (Spata (2006) – Access directive EC (2002d), ARt. 4(1), 5(1)).

[16] Formally, the procedure is called SSNIP (Small but significant and Non-transitory Increase in Price) test, which measures the pricing power of a (hypothetical) monopolist. Dobbs (2003) criticizes hypothetical monopolist tests such as SSNIP because they only rely on demand elasticities but ignore cost functions.

Performance Paradigm" from standard textbooks, adapted for the ex ante nature of regulation, should be used to measure effective competition.
3. Insufficiency of competition law, e.g. if the compliance requirements of an intervention ín a specific sector are extensive (Stumpf (2006)).

Having received the markets' recommendation lists from the European Commission (see Appendix 12.8), the national regulatory authorities are imposed to analyse the competitive situation in these markets, e.g. whether an operator has SMP. Briglauer (2007) points out that the first and second criteria of the three-criteria test above will also be taken into consideration in the SMP assessment process. Subsequently, the national regulator has to decide about remedies (see also Sect. 9.1.3). Furthermore, the national regulatory authorities could identify other markets in accordance with the three-criteria test. Regulatory decisions are made by the national regulatory authorities, but the European Commission has the right to file a veto. In antitrust cases, national agencies and courts are in charge as well as European institutions.

Regarding the international diffusion of networks, IT services and digital goods, there is a trade-off with respect to the definition of the relevant market: the larger a market is defined, the less probable it is to find SMP. But when misuse of SMP is existing, antitrust and/or regulatory measures could cumber the international diffusion of a company offering networks, IT services or digital goods. But on the other hand, firm-specific and overall effects may differ as short- and long-term effects do, because regulatory instruments (e.g. the obligation to grant access) may support (international) competitors.

8.1.7 Digitalization, Employment, and Education

Section 2.4 describes the development of productivity research: For a long time, till the 1990s, econometricians could not find empirical evidence for a positive influence of ICT investment on productivity, nor on sectoral or macroeconomic levels. This has changed after 2000: the results have indicated a positive relationship, although the relationship seems to become loose again now. For the additional problems which arise when digital goods are taken into consideration, see Sect. 2.4.4.

Just as the picture for productivity is not clear, it is as ambiguous for the effects of digitalization on employment. Generally, demand-side effects have also to be taken into account to determine the employment effect of productivity changes (Schettkat and Russo (2001)).

In political speeches, the labour-increasing effects of digitalization are usually emphasised, but from an economic point of view, digitalization can also be interpreted as a labour-saving innovation. The term "jobless growth", which was used in economic literature at the end of the 1990s (for example, in Bosch (1998)), describes the fear that ICT investments lead to productivity advances but not to a rise of employment.

On the other hand, mechanisms are at work which may compensate for the labour-saving effect. In their overview article, Spiezia and Vivarelli (2002) mention the following off-setting mechanisms for labour-saving innovations in the ICT sector:

- "Via new machines": New jobs are created in the capital sectors, which produce the machines for the ICT sector.
- "Via a decrease in prices": The unit costs of production decrease. Provided perfect competition is given, market prices decrease, which leads to a higher demand for products. Internationally, the competitiveness of a sector is driven by the introduction of cost-saving technologies, and employment compensation is obtained by an increase in world market shares.
- "Via new investments": Extra profits for innovative entrepreneurs may arise, even if there is a time lag between the implementation of cost-savings digital technologies and the decrease of prices in the good or service markets. These access profits may be invested further, and new jobs created.
- "Via a decrease in wages": A decrease in wages leads to an increase in the demand for labour, at least in a neo-classical framework.
- "Via an increase in income": Workers, maybe supported by labour unions, participate from higher productivity. Wages and incomes increase. The increase in income leads to higher demand and an increase in employment which compensates for initial job losses.
- "Via new products": New economic branches develop, which deploy new demand for labour.

The fourth mechanism, "decrease in wages", contradicts the fifth one, "increase in income", at least at first glance. Institutional frameworks, international competition, technology and market conditions in the labour and goods markets determine which effect dominates, and whether the labour-saving effect of digitalization can be compensated fully. Accordingly, the overall empirical picture is mixed, also because of a lot of difficulties of econometric research in this field (see also Sect. 2.4.3). But two results are commonly accepted:

1. Sector-level studies, which take into account the effects on rival companies, present evidence that process innovations have a stronger labour-displacing effect than product innovations (Spiezia and Vivarelli (2002)).
2. Firm-level studies show that investment in ICT changes the composition of labour demand: it diminishes the demand for low-skilled manual employment, raises the demand for skilled workers (Falk and Seim (2000), Autor et al. (2003), and Bartel et al. (2005)), and changes the international division of labour (Welfens (2007c)). A simple two-sector HOS-trade model makes clear the consequences: even if a sectoral shift occurs in only one of both countries - under the conditions of international factor price equalization[17] – the wedge between skilled and unskilled wages rises in both countries. Comparing the US and EU,

[17] The framework of a factor price equalization region is shown in Fig. 5.2.

Röger (2002) finds that a technology shock caused by ICT investments leads to rising skill premiums, but, in Europe, also to unemployment of the unskilled because of stronger labour market ridigities. However, Shaw (2002) points out that employment in the ICT sector (e.g. in a call centre) is not always skill-intensive. Welfens and Vogelsang (2008) argue that different IT services should be classified according to their knowledge intensity and economies of scale to evaluate their effects on growth, employment, and trade.

Confronted with these developments, it becomes clear that education and an appropriate education system are key factors for economic growth and employment (Welfens and Jungmittag (2002b)). Furthermore, governments try to promote digital technologies (see Sect. 9.2.4 for examples). An aspect might be the objective that their country get an advantage in international competition, but, as digital technology spreads worldwide, it is difficult to identify comparative advantages in the sense of Ricardo. Rather, the model presented in Chap. 4 illustrates that digitalization leads to higher export and import shares and a higher variety of goods. Furthermore, the international diffusion of networks and IT services helps to exploit existing differences in comparative advantages in traditional industries and services. In the literature about outsourcing, the consequences of digitalization are discussed extensively (see Amiti and Wei (2005), Feenstra (2007), and Sect. 3.3.1). Finally, it should be noted that "traditional" fiscal and monetary policies to enhance employment could become more difficult in the digital age, as digitalization leads to growing differences in adjustment speeds of goods, labour and financial markets, which favours international overshooting phenomena (Welfens (2001)), although countries maintain specific employment characteristics (Castells (2004)).

8.1.8 Trade Policy

8.1.8.1 Overview

In Sect. 3.3, the drivers for internationalization are described from the perspective of management and economic literature. As it has far-reaching effects on the economic structure of a country, governments started early to support or to dampen exports, imports, investments, or migration. Regarding trade, the most important instruments are:

- Customs duties or export subsidies
- Export or import quotas/contingents
- Non-tariff trade barriers, such as market entry conditions (e.g. setting specific standards).

The effects of these trade instruments depend on capital-labour endowments, market structures, technologies used, elasticities of substitution (demand), type of competition (e.g. Cournot or Bertrand competition in the case of an oligopoly), etc. For an overview, see Feenstra (2004) or Helpman and Krugman (1992).

The instruments mentioned directly influence the international diffusion of networks, IT services and digital goods. For example, Wiese (2002) shows that small tariffs may be quite effective in giving the domestic firm a decisive advantage, when network effects exist. On the other hand, there is strong support for the benefits of free trade (see Bhagwati (2007), for example). A strong argument against trade policy instruments is that measures of a country lead to retaliation, which counterbalances the terms-of-trade effects, but also impairs the diffusion of new technologies. To support the international development of e-commerce, WTO members agreed on a moratorium on e-commerce (see Sect. 8.2.1). Nevertheless, digitalization led to some new developments in the field of non-tariff trade barriers. Two examples:

- French law which may oblige US company Apple to make songs downloaded for iPod playable on other hardware[18] (for a discussion about intellectual property rights, see also Sect. 9.3.2).
- The attempts of China to control Internet content (Fowler (2008)) and the resulting self-censorship of search engine companies such as Google and Yahoo.[19]

The future relevance of such non-tariff trade barriers also depends on technological developments. On the one hand, technological solutions could be invented to bypass digital trade barriers. On the other, filters to control content and user access will also advance. They may be employed by countries receiving content (e.g. China) and countries sending content as well. For example, German public broadcaster ZDF restricted access to soccer broadcasts streamed via the Internet for clients in France during the World Championship 2006. However, from an economic point of view, the negative external effects of trade barriers for digital content for welfare and knowledge diffusion has to be weighed against the rights of intellectual property holders and their incentive to innovate.

Also, differences in tax regimes have an impact on trade, e.g. indirect taxes such as VAT (Value Added Tax) on e-commerce trade (GIIC (2000)). As a consumption tax, VAT should be collected in the country where the product is consumed, but in practice, governments collect taxes from the final seller of the product, assuming that consumption takes place close to the sale. But e-commerce in digital products "has blown appart this assumption", also because taxable online transactions are difficult to trace (Barfield (2001)).

8.1.8.2 Types of Trade Agreements

International rules on customs duties, quotas and market-entry conditions are laid down in bilateral, regional, and multilateral agreements between countries and

[18] Source: Financial Times Deutschland, 22 March 2006.

[19] Source: Financial Times Deutschland, 15 February 2006.

within the framework of the World Trade Organization (WTO). The trade agreements can be classified as below.

- *Bilateral agreements:* For example, the US Trade Representative (USTR (2006a)) lists 14 binding Free Trade Agreements (FTAs). Some of the FTAs also include special chapters on e-commerce; for an overview of these, see USTR (2006b). Wunsch-Vincent (2005a) compares the US bilateral trade agreements with Chile, Singapore and Australia, which include chapters on digital content products. They solve the service-or-good discussion (see Sect. 9.3.3) by introducing a "hyprid sector", confirm the applicability of WTO trade rules to e-commerce, and assure a zero-duty trade rate.
- *Multilateral agreements:* For example, the EC, US, Japan, Korea, and Taiwan negotiated an agreement to cut tariffs to zero on multi-chip integrated circuits (MCPs) in 2005. MCPs are high-technology semiconductors which wrap memory and non-memory chips into a single package. They are used primarily in smaller applications like cellphones, PDAs, and digital cameras (Portman (2005)). MCPs are not included in the ITA agreement (see Sect. 8.2.1) which was announced in 1996, because this type of chip had not been invented then. But if they had, it is likely that they would have been incorporated in ITA at zero duty (EC (2006b)). The duty-free treatment of MCPs is estimated to entail less customs revenue for the EC budget in the order of 7.9 million euro p.a. (EC (2006b)). The parties to the MCP agreement favour binding MCPs under the WTO later on. In this sense, the MCP agreement is an example for timing international trade policy: bi- and multilateral agreements pave the path for WTO coverage in compliance with the MFN principle[20] (UN (2002)).
- *Plurilateral agreements:* The most important are GATT, GATS, and TRIPS, which are part of the WTO system (and are described in Sect. 8.2.1 in detail).
- *Regional agreements:* As of January 2005, 170 Regional Trade Agreements are in force and have been notified to WTO. They are usually multilateral. Their greatest concentration is in Europe (Crawford and Fiorentino (2005)). The most important regional agreements are:
  - *EC/EU:* The EU has been a policy actor in digital markets and WTO negotiations (Sect. 8.2.2 is dedicated to the EU)
  - *EFTA (European Free Trade Association):* Members are Iceland, Liechtenstein, Norway and Switzerland
  - *NAFTA (North American Free Trade Agreement):* In 2005, goods and services in the amount of USD 810 billion were traded between the three NAFTA countries, the US, Mexico, and Canada (USTR (2006c))
  - *CAFTA-DR (Central America – Dominican Republic – United States Free Trade Agreement):* a free trade agreement between the US, Costa Rica, the Dominican Republic, El Salvador, Guatemala, Honduras, and Nicaragua. The US Congress and president approved the CAFTA-DR in the summer of 2005

[20] Most-favoured nation principle; see Sect. 8.2.1 for an explanation.

and it first entered into force between the US and El Salvador on 1 March 2006. CAFTA-DR determines the handling of digital goods explicitly: the members agree to not impose "duties, fees or other charges in connection with the importation or exportation of digital products by electronic transmission", and further declare that customs value of a digital product stored on a carrier medium should be only based on the cost of the carrier medium without regard to the value of the digital product itself (USTR (2005))
- *MERCOSUR:* A customs union between Argentina, Brazil, Paraguay and Uruguay
- *AFTA (Asian Free Trade Area):* Member states of AFTA and ASEAN (Association of Southeast Nations) are Brunei Darussalam, Indonesia, Malaysia, the Philippines, Singapore, and Thailand. AFTA was established in 2003. All members of AFTA are ASEAN member states also, but not vice versa (source: http://www.aseansec.org).

WTO agreements, which are agreed upon by all members, and bi- or multilateral agreements between some countries are in an ambivalent relationship to each other. On the one hand, Free Trade Agreements undermine the overarching claim and the Most Favored Nations Principle of the WTO. Therefore WTO staff use the term "Preferential Trade Agreement" (PTA) instead of "Free Trade Agreement" (FTA). On the other hand, FTAs are an instrument to liberalize trade further and to put pressure on WTO negotiation rounds. The dichotomous relationship was best expressed by the term "competitive liberalization", which was used by former US Trade Representative Zoellick when he announced a new trade strategy in February 2003 (Zoellick (2004)).

For example, in audiovisual sectors, countries go beyond their GATS offers, as traditionally, some countries have reservations against liberalized trading rules (see also Sects. 8.1.4 and 9.3.3). For example in the trade agreement between Hong Kong, China and Macao, China raised its foreign equity limit for movie projection and sound recording to 100 per cent. Trade agreements with the US trade in movie-related services are usually without limitations,[21] whereas with TV and radio, "a number of restrictions – including content quotas - are the norm rather than the exception", including foreign equity limitations for US radio owners. In contrast, "more liberal access is often granted where the US likely has more interests, e.g. satellite TV, foreign programming for cable TV, and interactive audio/video services" (Roy et al. (2006)).

For details about the impact of trade policy on the diffusion of IT services and digital goods, also see Sects. 9.2.3 and 9.3.3 respectively.

[21] Only Morocco, Colombia, Peru, the Dominican Republic, and Costa Rica have maintained market access restrictions in their agreements with the US (Roy et al. (2006))

8.2 Institutional Players

8.2.1 The WTO System

8.2.1.1 Overview

In March 2008, 151 countries were members of the World Trade Organisation[22]. The WTO negotiating programme is organized in rounds. The WTO was established on 1 January 1995 after the end of the so-called "Uruguay Round". The most recent negotiation round, the "Doha Round", started in November 2001 in Doha, Qatar. Since then, two ministerial meetings in Cancun, Mexico in September 2003 and Hong Kong in December 2004 have taken place. The initial timetable was planned to get ready with the Doha Round in 2005, but this timetable has been extended now.

WTO rules consider goods and services separately. The basic structures are presented in Table 8.3 and the subsequent sections.

8.2.1.2 GATT

The General Agreement on Tariffs and Trade (GATT) entered into force on 1 January 1948, in which 23 countries agreed "that their relations in the field of trade and economic endeavour should be conducted with a view to raising standards of living, ensuring full employment and a large and steadily growing volume of real income and effective demand, developing the full use of the resources of the world and expanding the production and exchange of goods, being desirous of contributing to these objectives by entering into reciprocal and mutually advantageous arrange-

Table 8.3 The basic structure of the WTO agreements, Source: WTO (2005d)

Umbrella	Agreement Establishing the WTO		
	Goods	Services	Intellectual Property
Basic principles	GATT	GATS	TRIPS
Additional details	Other goods agreements and annexes	Services annexes	
Market access commitments	Countries' schedules of commitments	Countries' schedules of commitments and MFN exemptions	
Dispute settlement	Dispute Settlement		
Transparency	Trade Policy Reviews		

[22] Source: http://www.wto.org.

ments directed to the substantial reduction of tariffs and other barriers to trade and to the elimination of discriminatory treatment in international commerce" (WTO (1986)). After the negotiations of the Uruguay Round, GATT acts as an "umbrella treaty" for trade in goods (WTO (2005d)). Specific sectors or issues are ruled upon in extra agreements and annexes. Finally, there are the detailed schedules of commitments made by individual countries. These take the form of binding commitments on tariffs or combinations of tariffs and quotas.[23] A sector-specific agreement for goods related to the digital economy is the ITA agreement, which is presented in Sect. 8.2.1.

8.2.1.3 GATS System

The General Agreement on Trade in Services (GATS) came into force in 1995 and refers to services. The principal provisions of GATS are as follows:

- General Obligations:
  - The Most-Favoured Nation (MFN) principle, which forbids WTO members from discriminating among their trading partners. If a country improves the conditions for one trading partner, it has to give the same treatment to imports of similar services and service suppliers from any other country. So all other WTO members remain "most-favoured" (WTO (2007b)). Exemptions to MFN are only allowed when they existed at the time of entry into force of the WTO Agreement.
  - The transparency obligation requires members to publish measures which affect trade in services, to inform the GATS council about them, to name contact points, etc.

  General obligations apply to all service sectors automatically, but exceptions can be recorded in the schedules of each WTO member (negative list approach). For example, in the revised offer from June 2005, the European Community lists an exception to the MFN principle with indefinite duration for the production and distribution of audiovisual works through broadcasting or other forms of transmission to the public (WTO (2005a)).
- Specific commitments:
  - *Market access:* When a member has committed itself to grant market access, it must not employ any restrictions such as monopolies, exclusive rights, quotas or tests on economic efficiency in this sector.
  - *National treatment principle:* It only applies when a service has been scheduled. Also, limitations can be inscribed in the schedule. The national treatment principle rules that a member must "accord to services and service suppliers

[23] After the Uruguay Round, individual countries' commitments on specific categories of goods and services were listed on 22,500 pages. Source: WTO (2005d).

of any other member...treatment not less favourable than that it accords to its own like services and service suppliers" (WTO (1994)).

Specific commitments are recorded in the schedules of each WTO member. They are relevant only when a member has explicity agreed on a commitment (positive list). Critics argue that market access and national treatment commitments alone are not sufficient to liberalize trade in services. Tough private data standards, stringent qualifications requirements, restrictions on outsourcing, etc. may work as non-tariff barriers and impact e-commerce (Wunsch-Vincent (2005b)).

The sector-specific commitments are scheduled for each mode of service delivery. GATS identifies four different modes (see figure): Mode 1 describes the cross-border supply of services similar to the delivery of goods. The supply of a service can be effected by telephone, fax, the Internet, TV, postal services or otherwise. When the customer (e.g. a tourist) moves to a foreign service supplier, this is classified as GATS mode 2 (consumption abroad). GATS mode 3 (commercial presence) describes the other way around, e.g. services delivered by a branch or subsidiary of a foreign-owned bank. Under GATS rules, this includes production, distribution, marketing and sales activities. Finally, when persons, self-employed or employees, move to a foreign country to supply services, this is part of GATS mode 4 (UN (2002)). Defining a mode also affects trade. For example, Sander (2002) points out that the liberalization impact would be largest íf e-commerce were filed under mode 2. Instead, mode 1 is preferred by consumer organizations, because then the transaction would be considered to have taken place in the buyer's country and its jurisdiction would apply.

GATS also has consequences for specific sectors. Krajewski (2005) analyses the impact of GATS on instruments to support art and culture:

- *Governments produce cultural services:* No contradiction with GATS rules. But accordant commitment for liberalization and privatization may possible.
- *Financial aid, fees, especially for public broadcast:* No contradiction with GATS rules when they are not discriminatory against foreign companies.
- *Quotas for TV and cinemas are a contradiction to general GATS rules:* Explicit exemption in commitments or no commitment at all necessary.

The WTO Annex on Telecommunications deals with access and use of public telecommunications networks and services. "An important number of WTO members" committed to open their markets to international competition, e.g. in removing market access barriers (Roy et al. (2006)). For suppliers of Internet services, this amounts to the right to interconnection (Bacchetta et al. (1998)). Almost all participants in the negotiations, which ended in 1997, made commitments in most sub-sectors. The markets of these participants account for more than 90% of global telecommunications revenues (Kathuria (2004)). Ten countries made explicit commitments on the supply of Internet access services in this context. But, on the other hand, the absence of such a commitment does not imply that market access opportunities for Internet access services do not exist.

Furthermore, in some bilateral agreements, the countries go beyond their GATS commitments, e.g. Panama committed to eliminate its telecommunications service monopoly in its trade agreement with El Salvador. Costa Rica committed to allow US telecommunications providers to supply direct private networks and Internet services. These two examples are directed to GATS mode 3 (direct investment). The cross-border supply (GATS mode 1) of basic telecommunications services is generally bound without limitations in trade agreements (Roy et al. (2006)).

Summarizing, services, such as telecommunications and computer-related services (see also Sect. 9.2.3), are covered by GATS. The picture is not clear for digital goods which can be delivered electronically (see Sect. 9.3.3).

8.2.1.4 Differences Between GATS and GATT

The important principles of both, GATT and GATS, are the Most-Favoured Nation (MFN) and the National Treatment principles. They are explained in the GATS Sect. 8.2.1 in detail. But GATT and GATS are different in the automatism applying the National Treatment principle. Bacchetta et al. (1998) mention the most important differences between GATS and GATT:

- GATT contains a general obligation in national treatment. But the GATS obligation in national treatment depends on the specific commitments that have been made by the WTO member states at sectoral level.
- GATT embodies a general prohibition on quantitative restrictions (with some exceptions), whereas GATS permits quantitative restrictions when governments wish to maintain limitations on market access.
- GATT only covers cross-border trade, whereas GATS also considers GATS modes 2, 3 and 4 (see Fig. 8.1) as part of services.

8.2.1.5 TRIPS Agreement and WIPO

The Agreement on Trade-Related Aspects of Intellectual Property Rights (TRIPS) covers copyright, trademark and patent issues and sets out minimum standards of protection to be provided by each member. TRIPS is an integral part of the WTO Agreement, and is binding for each member of the WTO.

Digitalization has two impacts on intellectual property rights (see also Sect. 9.3.2): The worldwide communication network changes the way protected products are sent and it is leading to new products which can be traded electronically (e.g. an mp3 file). The Internet, with its "borderless" nature, leads to difficulties in determining the applicability of territorially-based laws and regulations of activities carried out on a global network. WTO itself sees strong arguments to develop clear rules, because "copyright owners will be reluctant to put their protected material on the net as long as they fear that the Internet will lead to uncontrolled dissemination and copying of phonograms, films, computer programs and other protected material" (Bacchetta et al. (1998) – see also Sect. 2.3.3). Regarding digital technology,

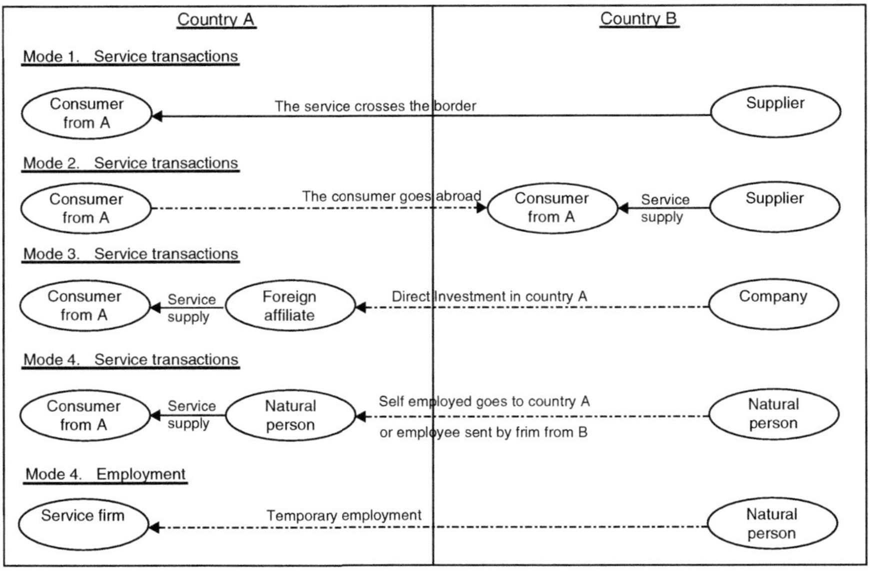

Fig. 8.1 Four modes of service delivery; Source: UN (2002)

the TRIPS agreement confirms that computer programmes must be protected as literary works, and clarifies that databases and other compilations of data or other material also have to be protected under copyright even if the databases include data that as such are not protected. For the relevance of the TRIPS agreement for the audiovisual sector, see Zampetti (2003).

In cooperation with WTO, the World Intellectual Property Organization (WIPO) has the task of implementing the TRIPS agreement. WIPO is part of the UN system of organizations and has 184 nations as member states (April 2008, source: www.wipo.org). It administers 23 international treaties dealing with different aspects of protecting intellectual property rights (IPR, see Sect. 9.3.2). Most important are the WIPO Copyright Treaty which requires the contracting parties to provide adequate and effective legal remedies against any person who, without authority, removes or alters digital watermarks, which are used to identify digital works. The WIPO Performances and Phonograms Treaty ensure authors, performers, and phonogram producers the exclusive right of authorizing the communication of how to make available to the public their protected material. These rights also cover on-demand delivery of works over the Internet. A recent topic of WIPO's "standing committee on copyright and related rights" is to develop a new treaty on the protection of broadcasting organizations. WIPO argues that also "the growing signal piracy, including digitalized pre-broadcast signals has created a need to update" current rules (WIPO (2005)).

8.2.1.6 ITA

The Information Technology Agreement (ITA) which is a plurilateral agreement on MFN basis, was signed by 28 WTO members in Singapore in December 1996. The signatories to ITA must eliminate their tariffs on the ITA products for all WTO members, even though the latter may not have signed the ITA. This may lead to a potential "free-rider" problem (Wunsch-Vincent (2005b)). But the distortions may remain small, because all important countries trading IT products have signed the agreement.[24]

In the ministerial declaration, it is stated that the elimination of customs duties should take place in four stages, but in no case later than 2005. The product coverage under ITA include semiconductors and semiconductor manufacturing equipment, telephone sets, radio and TV transmission apparatus, pagers, fax machines, switches, modems, scientific instruments such as cromatographs and spectrometers, computer hardware such as processing units, keyboards, printers, monitors, and storage media such as diskettes, CDs.

Also, non-tariff measures such as national certifications requirements have been addressed by ITA (Wunsch-Vincent (2005b)), but they don't play a special role during the Doha Round.

Due to technological improvements, newer products are not covered by ITA. An example are multi-chips integrated circuits (MCP). In 2005, the EC, USA, Japan, Korea, and Taiwan negotiated the MCP Agreement to cut tariffs to zero on MCPs (EC (2006b), Portman (2005)).

8.2.1.7 Moratorium on E-Commerce

In November 2001, the Fourth Ministerial Conference in Doha, Qatar, provided the mandate for a work programme on e-commerce to clarify the classification issues (see Sect. 9.3.3) and to integrate rules into the existing WTO systems. A moratorium on customs duties was concluded which was prolonged at the the most recent meeting of trade ministers in Hong Kong in December 2005. The declarations states: "E-commerce: We take note ... that the examination of issues under the Work Programme is not yet complete. We agree to reinvigorate that work, including...discussions on the trade treatment, inter alia, of electronically delivered software...We declare that Members will maintain their current practice of not

[24] These are Albania, Australia, Austria, Bahrain, Belgium, Bulgaria, Canada, China, Costa Rica, Croatia, Cyprus, The Czech Republic, Denmark, Egypt, El Salvador, Estonia, The European Communities (which count 15 member state), France, Finland, Georgia, Germany, Greece, Hong Kong, Hungary, Iceland, India, Indonesia, Ireland, Israel, Italy, Japan, Jordan, Korea, The Kyrgyz Republic, Latvia, Liechtenstein, Lithuania, Luxembourg, Macao, Malaysia, Malta, Mauritius, Moldova, Morocco, The Netherlands, New Zealand, Norway, Oman, Panama, The Philippines, Poland, Portugal, Romania, Singapore, The Slovak Republic, Slovenia, Spain, Sweden, Switzerland, Chinese Taipei, Thailand, Turkey, UK, US (Source: http://www.wto.org/english/tratop_e/inftec_e/itapart_e.htm, accessed March 2006).

imposing customs duties on electronic transmissions until our next Session" (source: WTO (2005b)).

The moratorium is more a political commitment by the WTO members than a binding rule. It cannot be legally enforced via the WTO dispute settlement system (see Sect. 8.2.1). Also, it remains unclear what "electronic transmissions" are: only the transport services, or also the electronically-delivered content? Furthermore, there is criticism that the moratorium is not technologically neutral when the same content delivered on CD may be subject to customs duties. Also, developed countries, which rely on taxes, may benefit more than developing countries, which depend more on revenues from customs duties (Wunsch-Vincent (2005b)).

8.2.1.8 GPA

According to the Agreement on Government Procurement (GPA), the signatories should ensure that the use of IT promotes the aims of an open, non-discriminatory and efficient government procurement through transparent procedures.

8.2.1.9 Dispute Settlement

When a WTO member believes that another member violates its general obligations or specific commitments, it can call the Dispute Settlement Body. For example, in a trade dispute between Antigua and the US on market access for gambling and betting on Internet services in the US, the WTO arbitrator ruled that the US would have to open this market till April 2006 (WTO (2005e)).

8.2.1.10 Summary of WTO System

Table 8.4 provides an overview about the agreements and the products covered.

For bi- or multilateral trade agreements outside the WTO framework (such as the MCP Agreement for multi-chip integrated circuits), see Sect. 8.1.8.

8.2.2 The European Community Approach

At the European level, two different contractual frameworks have to be differentiated: the European Union (EU), and the European Community (EC). This differentiation between the EU and EC is not important for the further parts of this thesis but needs to be mentioned here. The "Treaty Establishing the European Union" (EC (2002c)) deals with foreign and security policy and the Atomic Energy Community only. The rights of member states and institutions differ in EU and EC frameworks. For example, Art. 21 of the EU Treaty rules that, "the Presidency (of

Table 8.4 International trade rules and digital products

Rule/Agreement	Topic	Remarks
GATT	Goods in general	Also CDs, DVDs with software and other content
GATS	All Services, also telecommunications, internet infrastructure and computer related services; sector-specific commitments necessary	Only a small number of countries have made commitments to audio-visual services; Open: Classify digitally delivered products, which have traditionally been traded on a physical carrier medium, as goods (GATT) or services (GATS)?
GATS Telecommunication Annex	For Internet Service Suppliers: Interconnection Right	
ITA	Infrastructure Goods such as semiconductors, computers etc.	tariffs at zero
e-commerce Moratorium	No customs duties on electronic transmissions	Moratorium prolonged until next Ministerial Conference. Rather political commitment than binding rule.
GPA-Agreement on Goverment Procurement	Government should use IT for transparent procurement procedures	
TRIPS Agreement	Intellectual Property Rights	

the council) shall consult the European Parliament on the main aspects and the basic choices of the common foreign and security policy and shall ensure that the views of the European Parliament (EP) are duly taken into consideration" (EC (2002c)), whereas the EC gives more rights to the EP, at least in some political fields.

The goal of the EC is a common economic and social policy. The Treaty Establishing the European Community (EC (2002b)) defines the intention and instruments. One characteristic is the shared competence between the EC and the member states. This may also lead to problems for the contract parties of international economic agreements (Leal-Arcas (2003)).

The next institutional reform of the EU/EC should be the Treaty of Lisbon. It was signed by the political leaders of the 27 member states in Lisbon in December 2007, but has not been put in force yet due to the pending ratification process.

Article 249 of the EC treaty (EC (2002b)) mentions the most important instruments to develop European 'legislation' currently:

- "*A regulation* shall have general application. It shall be binding in its entirety and directly applicable in all Member States.
- *A directive* shall be binding, as to the result to be achieved, upon each Member State to which it is addressed, but shall leave to the national authorities the choice of form and methods.

- *A decision* shall be binding in its entirety upon those to whom it is addressed.
- *Recommendations and opinions* shall have no binding force."

Regarding the topic of digitalization, the most important directive is the framework "Directive for electronic communications networks and services" (2002/21/EC, see EC (2002e)). It outlines general principles and objectives on all topics in this field, including broadcasting and telecommunications. This framework directive is supplemented by more specific rules which are related to specific topics of electronic communications and/or sector-specific rules: "Directive (2002/19/EC) on access and interconnection, Directive (2002/20/EC) on the authorisation of electronic communications networks and services, Directive (2002/22/EC) on universal service and users" rights, Directive (2002/58/EC) on privacy and electronic communications, Directive (2002/77/EC) on competition in the markets for electronic communications services, Regulation (2000/2887/EC) on unbundled access to the local loop.[25] The framework does not address content. The "Audiovisual Media Services Directive" (Directive 2007/65/EC) is specific for TV and related services. A characteristic of European legislation is the step-by-step implementation of European rules, e.g. the "Audiovisual Media Services Directive" amends the "Television without Frontiers Directive" and renames it. Member states have to work national laws over till 2009 accordingly.[26]

But de facto, the actual legal framework which is relevant for companies and consumers is not only developed by the European Commission, European Parliament and European Courts, but also by national parliaments, courts and regulatory agencies. Additionally, special bodies such as the European Regulators Group (ERG) which is based on the membership of all National Regulatory Authorities (NRAs) foster the harmonization process between national application of European guidelines. For more information[27] about European regulatory policy, also see Sect. 8.1.6. For a commendation of the EU's framework, see Mindel and Sicker (2006) and Sect. 10.2.

8.2.3 Internet Governance

The different layers shown in Table 2.1 have to interoperate for the Internet to work. For example, the protocols and procedures to transfer www.uni-wuppertal.de into the IP address 132.195.129.27 have to be defined. This is the task of "Internet gover-

[25] Source: http://ec.europa.eu/information_society/policy/ecomm/current/index_en.htm.

[26] Sources: http://ec.europa.eu/avpolicy/reg/tvwf/index_en.htm and http://ec.europa.eu/avpolicy/reg/avms/index_en.htm.

[27] A comment: Alexander Streel presents an illustrative figure in a book of Justus Haucap and Ralf Dewenter (eds), published by Elsevier in 2007. Unfortunately Elsevier has sold the rights to a third party company, which does not accept the reuse of the figure for scientific purposes without being paid for.

nance". According to Eric and Curien (2007), the three most important organizations which guarantee (inter)operability of the Internet are the following:

- *ICANN: Internet Corporation for Assigned Names and Numbers:* Under a contract with the US Department of Commerce, ICANN is responsible for organizing the distribution of IP numbers and domain names. ICANN sets the available first-order suffixes (Top Level Domains – TLD) such as .com, .de, .it, etc. It is criticized that US government could, theoretically, put pressure on ICANN to switch off a TLD and ICANN exploits its monopoly by a shortage of TLDs. According to Müller et al. (2006) the best solution would be a divesture of ICANN and a new foundation at international level. On the other hand Christou and Simpson (2007) argue that there were no realitistic structural alternatives to ICANN and show how the European Union has achieved more direct governmental influence in ICANN, not least due to "rhetoric action".
- *IETF: Internet Engineering Task Force:* IETF is de facto responsible for the standardization of the communication protocols.
- *W3C: World Wide Web Consortium (W3C):* The W3C decides about the standardization of multimedia languages used. It is a kind of a club, as companies can become members.[28] For criticism of this approach, see the quote at the end of Sect. 9.2.2.

All three institutions are subject to US law. Other organizations such as the UN prefer a "multi-stakeholder approach", e.g. during the 2005 World Summit on the Information Society (WSIS) in Tunis, it was emphasised that this was necessary to develop a global information society and to take coordinated actions against threats such as spam, viruses, etc. (ITU and UNCTAD (2007)).

---

[28] The member list shows 421 members in April 2008. Source: http://www.w3.org/Consortium/Member/List.

Chapter 9
Policy for Networks, Services, Digital Goods

9.1 Networks

9.1.1 Political Goals

Data networks are necessary to have access to digital goods and IT services. Different types of networks (fixed line, mobile, etc.) are described in Sect. 2.3.1. Macroeconomic consequences of a spread of networks are described by the model in Chap. 4. The basic model of network economics which emphasises network effects on the demand side is presented in Sect. 3.1.1. A combination of network effects, lock-in effects, sunk costs, economies of scale and scope may lead to a monopolization of network supply. Different kinds of networks (energy, telephone, water, etc.) are also labeled as "natural monopolies" in standard microeconomic theory because of continuously-falling average costs when demand increases.

The liberalization of telecommunications networks in Europe has shown that network monopolies can be opened by price and access regulation. The general goals of regulatory economic policy vary. In literature, it is claimed that network access should be provided:

- Efficiently (for productive, allocative and dynamic efficiency (see Sect. 8.1.2))
- With high quality and low prices (competitive)
- Universally to all groups of users
- Inducing the development of new products fostering economic growth (Welfens (2006)).

Broadband policies (see also OECD (2006a)) can include special broadband policies for regions (e.g. rural and remote areas), users (e.g. schools, universities, etc.) or technologies (wireless, cable, etc). Besides competition and regulatory policies, further instruments may be direct funding, tax policies, subsidies, promotion initiatives, or progressive e-government solutions as technological drivers. An overview of several national and regional (EU) broadband action plans is given by OECD (2007e). In the following section, I focus on the market structure and regulatory policy options of the Internet backbone and access market.

M. Vogelsang, *Digitalization in Open Economies*, Contributions to Economics,
DOI 10.1007/978-3-7908-2392-9_9, © Springer-Verlag Berlin Heidelberg 2010

9.1.2 Access Prices and Interconnection

Figure 9.1 shows a data network topology (see also Sect. 2.2.1) and the several ways an IP traffic packet can follow from the web server to a user. This figure represents the physical layer (see Table 2.1) of a network, the cables and hardware. Note that the logical layer which encompasses the administration of the customer account, log-in procedure, routing, etc. can be separated from the physical layer. For example, when an ISP offers an Internet dial-in service over a telephone line, it does not have be the owner of the last mile.

The interconnection points are provided by IXPs (Internet exchange points). They are "neutral" and generally non-profit where the members of the IXP exchange traffic (OECD (2006d)). Regulatory questions can arise on IXPs as well as at the edges of the network:

- *The interconnection* of providers, especially between Internet Service Providers (ISPs) and Internet Backbone Providers (IBPs) or among IBPs ("Wholesale business" in the middle of Fig. 9.1). Interconnection can be differentiated further in peering and transit (see Sect. 2.3.1). The rationale to peer is to avoid transit fees and to diminish latency times for data requests from other networks.[1] The

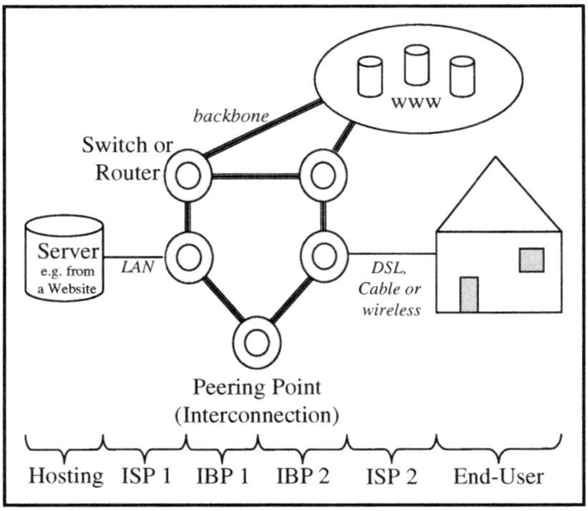

Fig. 9.1 Structure of internet routing

[1] De-peering can also occur. OECD (2006d) reports that TelstaClear de-peered at the Wellington, New Zealand IXP (WIX) in 2005. So TelstraClear's customers could still reach sites from TradeMe (a favourite auction site in New Zealand) or the Streams of Radio New Zealand, but the traffic was routed through international networks. This leads to higher latency times and that only parts of the programmes of Radio New Zealand (which is restricted for international listeners due to copyright reasons) could be received by customers of TelstraClear. TelstraClear hoped the content provider would become paying transit customers.

procedure to find an optimal set of peering of transit partners is discussed in business literature; see Heckmann et al. (2004) for an example.
- *The access* to the network.
  - *Retail-level/end-user-connection:* In the case of the Internet, these are the conditions to access via a dial-in connection, leased line, DSL, cable and/or wireless.
  - *Access of the content provider (Hosting):* The connection of servers to the Internet. The servers (e.g. a web server) deliver the requested data (e.g. a web page) to several users at one point of time.

  Access for both ends of the network as shown in Fig. 9.1 are provided by ISPs.[2] Although in the broadband age, ISP services are also offered by IBPs or their affiliated sister companies directly (Nuechterlein and Weiser (2005)). The traditional separation between IBPs and ISPs stems from former times, when dial-in services via a modem were in vogue.

9.1.2.1 Lessons from the Telecommunications Sector

There is no uniform use of terminology, as strands of technological and economics literature meet here.[3] In papers about regulation in the telecommunications sector[4] authors usually differentiate between "one-way-access" and "two-way-access". One-way access refers to a monopoly input: other firms need access to the incumbent's bottleneck facility to be able to compete at the retail level. With "two-way access" (or horizontal interconnection), competing firms grant access to each other's essential facilities respecting each other's networks. Sometimes (e.g. Dewenter and Haucap (2007)), "access" is also used as a generic term for all types of access and interconnection.

---

[2] Ofcom (2007a) differentiates between "connections", "hosting" and "other services" of an ISP. "Other services" are domain name registration, consultancy, website design, etc.

[3] More linguistic confusion may arise because of different use of "downstream" and "upstream". In information technology "downstream" and "upstream" change according to the function of the IP packets (e.g. the request for an email download is "upstream", whereas the download of emails is "downstream"). In economics, the connection from the ISP to the retailer is called "downstream", the connection from the ISP to IBP is termed "upstream".

[4] The trade-offs between the goals and the several instruments of regulation of telecommunications networks are described by a vast amount of literature, e.g. as an overview: Economides (2004) and Tardiff (2006); more theoretically, Armstrong (2002); comparing the EU and US approaches, Geradin and Sidak (2005) and Vanberg (2007); focusing on Germany, Dewenter and Haucap (2004); or suggesting new regulation approaches, Welfens (2007b). Although competition in telecommunications has increased and new services lead to a shrinking of the monopolistic bottleneck within the local loop (Blankert et al. (2006)), a warning on the danger of a squeezing out also in liberalized markets remains necessary according to Haucap and Kruse (2004).

Access and interconnection could be subject to monopolistic bottlenecks, e.g. caused by property rights of the essential facilities[5] (the last mile), and lock-in-effects, economies of scale or scope. Vanberg (2005) claims that "only when a network operator already has network specific market power due to the ownership of a monopolistic bottleneck area, will network externalities enable the operator to increase his market dominance". On the other hand, in the case of a contestable natural monopoly also, a new entrant will need to gain a critical mass of users first.

In the telecommunications sector, the incumbents have owned the networks in most countries. The theory of telecom regulation emphasises the impact of access and interconnection prices on retail prices, and, with it, on consumers' utility function. In the easiest models of one-way access, the retail prices of the incumbent and new entrant are equal, but the incumbent sets the access fee for the new competitor to its network. Without price regulation, the incumbent could squeeze out the entrant by high access and low retail prices, which amounts to a foreclosure strategy.

In models of two-way access, some kind of inertia of consumers is usually assumed, e.g. local markets or product differentiation (Kocsis (2007)). When the propensity to make telephone calls of a specific user group is higher than for another group, then it is profit-maximizing for duopolists to set high access prices and to offer their services only to a segment of the potential customers (Shy (2002)).

Otherwise, all totally elastic demand would be allotted to the company which sets its retail price marginally lower (Bertrand competition), or a symmetric solution must arise. For example, Armstrong (1998) analyses a two-stage game with two competing companies having local loops and long-distance networks. Users can call users of the competing network. Both providers set access fees and retail prices. In equilibrium, the duopoly behaves symmetrically and the firms have no incentive to deviate from the collusive retail price, even though they are set non-cooperatively. Here, collusion occurs, because the competing networks can directly influence the other networks' (perceived) costs by increasing their charges (Dewenter and Haucap (2007)). Vogelsang (2003) points out that collusion is an issue in symmetric interconnection relationships, but also exclusion is likely when a large network provider interconnects with a small network provider.

9.1.2.2 Theory of Interconnection Applied on IP-Based Connections

Aside from pricing, the historical experience with interconnection in the telecommunications sector shows that interconnection was not made available freely by an incumbent to its competitors. But in contrast to this, the (largely unregulated) internet service providers provide interconnection to each other voluntarily (Noam (2002)). The reasons for the differences between the sectors are as follows:

---

[5] The FCC, the US telecommunications regulator, uses the "essential facility doctrine" to decide on regulatory remedies. This approach is different from the EU's "significant market power" test. For similarities and differences between the approaches, see Vanberg (2007) and Geradin and Sidak (2005).

- The "competitive" structure of packet-switched routing (see Sect. 2.2.1). An IP packet can use many different ways to reach the final destination (see also Fig. 9.1). This lowers the market power of incumbents. For example, Giovannetti (2004) reports that in the 1990s, many European ISPs had leased bandwidth in the US to route intra-European traffic due to differences in transit prices.
- (Large) Companies and institutions can directly connect to network providers at IXPs.[6]
- Customers demand "universal connectivity" from their ISP because they do not know in advance what web page they may want to visit or whom they may want to send emails to (Economides (2005)). Providers realize the positive network externalities to the users in having a larger number of network participants ((Noam (2002))).
- In the Internet, the intelligence lies at the edges of the network. For the consequences of this "end-to-end-principle" and network neutrality, see Sect. 9.1.5.
- The protocols to address and route the data packets are public and not a proprietary standard.[7]
- Destination-related price discrimination is practically impossible in the Internet (Jahn and Prüfer (2006)), although it is usual in telecommunications (termination fee).
- In the Internet, receiving traffic is also accounted for in volume-based tariffs. In telecommunications, the receiver does not usually pay for receiving calls (Laffont et al. (2003)). Cross-border mobile calls and receiver party-paid regimes as in Russia are exceptions.

Summarising, the Internet is not expected to "leverage on network externalities and create a bottleneck" (Economides (2005)). The small market shares in peers which are shown in Fig. 2.3 promote this finding. On the other hand, theory predicts that in some cases ("one-way structure") regulation might also become necessary with interconnection of ISPs and IBPs, e.g. a model similar to the approach of the telecommunications sector.[8] Ida and Ueda (2004) consider a duopolistic component pricing model, where the server of an ISP (hosting provider) has to be connected to the World Wide Web. Assuming a "system" consists of two elements, interconnection and access services. If the incumbent offers IBP and ISP services, but a new entrant can only supply ISP services (one-way competition), the incumbent will set the price of its ISP services below the price of its competitor, because it takes system

[6] These companies act as Autonomous Systems (AS). An AS "is one or more IP networks, under the same management, which have defined and coherent external routing policy" (OECD (2006d), e.g. Google has the Autonomous System Number 15169.

[7] For example, TCP/IP (Transmission Control Protocol / Internet Protocol) or the Domain Name System (DNS) relate domains like uni-wuppertal.de to IP addresses. As IP addresses become rare, a new standard (IPv6) is developed.

[8] Ida and Ueda (2004) use demand functions instead of utility functions and allow any combination of "systems" with two-way competition, e.g. ISP services of company i with with IBP services of company j, with $i, j \in \{1, 2\}$

demand into consideration, not only component demand. The incumbent prices the system and regulatory instruments (see Sect. 9.1.3) should be used.

Dewenter and Haucap (2007) point out that foreclosure of the new competitors is the main risk associated with one-way access, although they admit that foreclosure can also happen with two-way access, especially when the networks are of different sizes. In the symmetric set-up of two-way competition in the model of Ida and Ueda (2004) (both companies are offering interconnection and access services), prices and demand are also symmetric and no regulatory efforts seem necessary. This is also true in the model of Laffont et al. (2003) with consumers and content providers. They show that the access charge negotiated between two rival network providers merely determines how the cost of traffic is shared between senders (websites) and receivers (end-users). In this model with perfect Bertrand competition in the two-sided backbone market prices of the two rivals are symmetric and their profits amount to zero. Giovannetti (2005) emphasises the importance of consumer behaviour. In a price-competition model of three different Internet providers in two regions, she analyses the effects when an alternative routing path between the regions becomes available (interconnection). She shows that interconnection lowers retail prices only if consumers are indifferent about the possible ways of accessing the Internet.

From the economic literature, the policy conclusions for the interconnection market are ambiguous. The consensus is that regimes in wholesale and retail markets are interdependent. But the actual effects on welfare and the need for economic policy to regulate the market depend on the factors mentioned in the first row of Table 9.1. Furthermore, Jahn and Prüfer (2006) point to national interests: they differentiate between US- and non-US politicians, because all top-level backbones are US-based and non-US policy-makers do not include profits from IP transit in their calculus.

Table 9.1 Internet interconnection models and policy recommendations

Theoretical focus	Emphasised by	Policy recommendation
One-way or two-way competition between IBPs and ISPs	Ida and Ueda (2004)	Regulation if one-way-competition
Product differentiation between ways of network access/love of variety of consumers	Giovannetti (2005)	Promoting interconnection if high competition at retail level
Size of competing networks	Jahn and Prüfer (2006)	Non-US-politicians: promoting competition between intermediaries (transit providers)
Timing of setting wholesale and retail prices	esp. Giovannetti (2005)	Prices are set simultaneously when there is no market power for IBP

9.1.3 Regulatory Instruments

Access and interconnection to networks can be subject to regulation. In general, the instrument are:

- Non-pricing remedies:
  - Disintegration or functional separation of a dominating company; the latter implies separation of operations (e.g. the support centre), employees (e.g. physically separate offices), and information (so-called "chinese walls") (ERG (2007b))
  - Unbundling/lease line
  - Obligation to grant access/"multiple access"
  - Net neutrality (see Sect. 9.1.5)
- Pricing rules (see Appendix 12.9 for an overview).

In telecommunications, the most important non-pricing rule is the obligation to grant access. In the case of the last mile to the consumer, the owner of the cable (e.g. an incumbent) has to lease the line (Local Loop Unbundling).[9] Or, it grants Internet access only (bitstream access[10]) to competitors.

In the Internet, monopolistic market structures may occur at the edges of the network in countries with a singular backbone provider through the convergence of services or due to mergers. But generally monopolistic bottlenecks are not expected with regard to Internet interconnections (Economides (2005)). In contrast to the ISPs at the "edges" of the network (see Fig. 9.1), the IBPs have peer to each other at zero charge (bill-and-keep) initially. But this has also led to the strategy that each provider tried to unload its traffic as quickly as possible to other and "to provide them with poor services to discourage them from doing the same" (Noam (2002)). The zero-charge tradition may lead to imbalances between providers which connect different types of content.

But if regulating prices of access and interconnection seems to be necessary (e.g. in countries with a dominating IBP), the price instruments discussed in the telecommunications sector could be applied, although the warning of Giovannetti (2004) should be taken into account: in the Internet, packets are not routed along a dedicated line as in circuit-switched networks, but packet switching is used instead. This also changes the effects of remedies applied.

---

[9] Definition used by Ofcom (2007b): LLU is the process whereby incumbent operators (in the UK this means BT and Kingston Communications) make their local network (the lines that run from the customer's premises to the telephone exchange) available to other communications providers. The process requires the competitor to deploy its own equipment in the incumbent's local exchange and to establish a backhaul connection between this equipment and its core network.

[10] Definition used by Ofcom (2007b): Bitstream is a wholesale service providing conveyance of data traffic from an end-user's premise to a point of interconnection made available by the incumbent to a competitive provider.

9.1.4 New Generation Networks

New Generation Networks (NGN) are broadband networks which are built to transport all kinds of communication data. TV, IP and PSTN networks converge. For a precise definition of NGN, see Sect. 2.3.1. The foreseen spread of NGNs also has impacts on regulatory aspects when former separated networks migrate to a "universal" network. TV and PSTN networks are usually regulated; IP networks (at least at the wholesale level) are not. Table 9.5 summarises technological origins and the most important regulatory principles of the different networks.

Which regulatory regime, if any, should be applied in an NGN-world is an open question. The following effects of NGNs have to be taken into account:

- The competition at the retail level increases as network access is possible via cable (TV) or wireless
- The relatedness of the markets increases (e.g. voice applications will be implemented on Web pages)[11]
- Network access and services can be separated in NGNs[12]
- With interconnection, the interoperability of IP-based services also becomes an issue (ERG (2007d)).
- Termination payments which are usual in the telecommunications sector may fall apart in NGNs. So a "funding vehicle" for universal services (in rural parts, developing countries, etc.) may also disappear (ITU (2006b)). For the discussion to extend the universal service (financing) obligation to broadband, see OECD (2006g)
- With NGNs, investment decisions come into the focus of the regulatory debate

NGNs rely on new cables (fibres) and hardware (switches, routers, etc.) or wireless infrastructure. The announced regulatory regime is the main determinant for the expected profit of the potential investors and determines the market structure ex ante. ERG (2007c) points out that different NGN interconnection points[13] also influence which remedies might be possible. The possibilities of regulating evolving NGNs are generally as follows:

1. "Regulatory holidays", e.g. means the promise not to regulate, either wholesale or retail market or both
2. Regulatory regime ex ante

[11] Auction company eBay acquired VoIP provider Skype in 2005.

[12] Nuechterlein and Weiser (2005) observe that the former separation between IBPs and ISPs has already vanished. Physical layer (cable) and logical layer (log-in, routing, etc.) are offered by one company or the affiliated ISP of an IBP in the US broadband business.

[13] It can be differentiated between FttCab (Fibre to the Cabinet, copper to the home); FttB (Fibre to the building, copper to the apartments); and FttH (Fibre to the home). With decreasing copper loop length, the available bandwidth increases. If FttH is built up with a shared infrastructure (point-to-multipoint), unbundling would be a technical problem (ERG (2007c)).

(a) Licensing, especially with regard to the distribution of new wireless frequency spectra
(b) Pricing and access: participants can build their expectations on the basis of the announced regulation principles

3. Regulation ex post: regulation according to the resulting market structure.

A prerequisite for the development of new (voice) services within NGN is that the operators can interconnect to both existing Internet exchanges and PSTN (e.g. OECD (2006c), which analyses the implication of WiMax technology). A special issue is the migration from PSTN to NGN by the telecommunications provider. On the one hand, the use of old copper wires for the last mile limits the available bandwidth, as Fig. 9.2 shows (explanation: see footnote).

Using the "old" copper cable which is often the property of the ex-monopolist in many countries also means to perpetuate the bottleneck problem. On the other hand, it is discussed which regulatory remedies (if any) should be used for interconnection. Instruments for access and price regulation are listed in Sect. 9.1.3 and Appendix 12.9. If price regulation is necessary, minutes or bandwidth (capacity) can be billed. For a general overview of this topic, see Vogelsang (2006), Marcus (2006) and Kocsis and Bijl (2007). Berg et al. (2006) point out that many details are not clear, e.g. the number of interconnection points necessary in future, although the process of migration has already begun. ERG (2007c) stress that the European regulators are very ancient in that telecommunications incumbents extend market

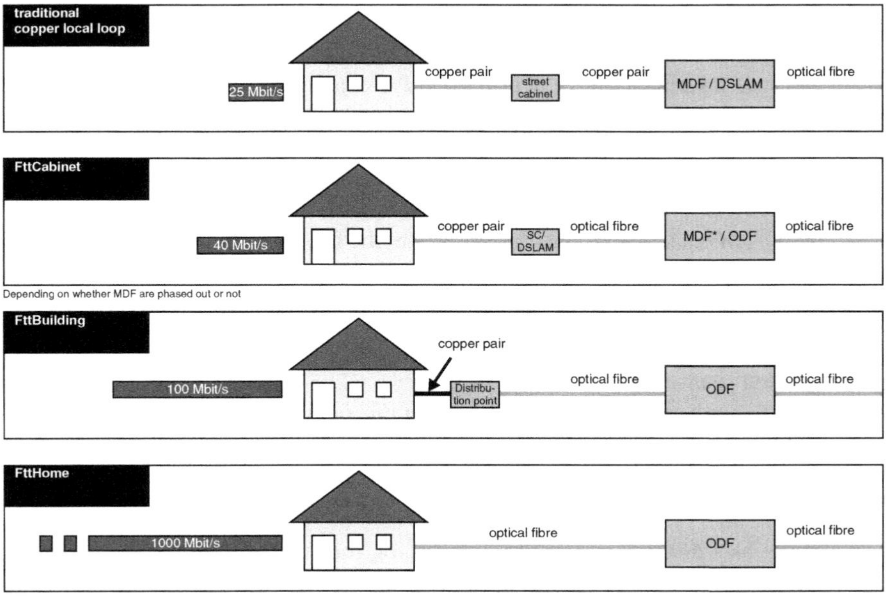

Fig. 9.2 Access architectures and typical maximum speeds. Source: ERG (2007a), p.66, based on an illustration of ARCEP, the French Regulation Authority

power when migrating from PSTN to NGN networks. Other authors emphasise the role of innovations which are fostered by (temporary) monopoly rents. For example, Welfens (2006) recommends suspending the regulation of the new fibre network in Germany for at least 1 year.

9.1.5 Network Neutrality

The "intelligence" of a data network resides at the periphery of the network shown in Fig. 9.1. The "intelligence" consists of the server software and the application software (e.g. the browser, the video client), which interact. The network in between is the "dump" in the first instance. Lessig considers this a defining principle of the Internet in contrast to telecommunications networks: applications were run on computers at the edge of the network. So innovaters with new applications needed only to connect their computers to the network. "Like a daydreaming postal worker, the network simply moves the data and leaves interpretation of the data to the applications at either end" (Lessig (1999), p.32).

Following this "end-to-end-principle" leads to the demand of "network neutrality". A network is called neutral when "it does not distinguish between the applications, or content, that depend on it, nor on the identity or nature of the users" (UNCTAD (2006)). The four non-discrimination principles from the FCC (see Sect. 8.1.4) are a manifestation of network neutrality.

A proposal is that governments should obligate all ISPs to provide Internet services consistent with the end-to-end-principle. This would be a more direct regulatory strategy than unbundling services "if the concern at stake is that network providers will leverage control over the network into some control over content". Nuechterlein and Weiser (2005) recommend establishing network neutrality as a main regulatory principle even for broadband networks.

But ISPs de facto can give priority to selected data packets, that means they can deploy content-discriminating technologies. This can be related to the sender of the data, the type of data, or a request for priority in the header of the data packet. Today discrimination is applied:

- By time-critical applications (e.g. VoIP or gaming)
- For security reasons (e.g. to protect children, specific content may be blocked)
- To prevent spam
- To comply with copyright law (see the example of New Zealand Radio in the first footnote of Sect. 9.1.2)
- To censor content (e.g. for the recent attempts of China to control Internet content see Fowler (2008)).

Gupta et al. (2005) argue that proper pricing mechanisms will be necessary to manage quality of data transport in future. Otherwise it would be "impossible to allocate network resources efficiently" because the demand for quality of service

(QoS) of Internet connections will increase due to new applications[14] despite faster infrastructure and new technologies to control network traffic. OECD (2007c) concludes that "the ability to designate priority to certain applications will be a boon for consumers and providers as long as there is sufficient competition in the market". Kocsis and Bijl (2007) agree with this, but also argue that it is hard to detect when a network operator gives bandwidth priorities to specific applications, services, or content providers.

9.2 IT Services

9.2.1 *Goals*

The aims of economic policy related to IT services are more general than for the network sector. In contrast to networks, no physical bottlenecks ("last mile") exist here, which are the starting point for regulation of networks and the corresponding economic literature. Also, gaining significant market power is more difficult with services. They cannot be stored so they are tailored more specifically to consumers as is argued in Sect. 3.1.2.

Policies directly related to IT services focus on the following:

- Promoting domestic IT services internationally. This includes outsourcing/offshoring services (see Sects. 2.3.2 and 3.3.1). The general equilibrium model of Chap. 5 shows that IT services are traded internationally even if there are no differences in factor prices or technology. Standard-setting (Sect. 9.2.2) and the negotiation of trading rules (Sect. 9.2.3) are important economic instruments to promote domestic IT service providers.
- Promoting IT services domestically to raise consumer welfare/productivity of government or private business (Sect. 9.2.4).
- Security aspects, e.g. in Germany, the email provider has to offer interfaces to security authorities so that the latter can get copies of emails easily also.

More generally, Arora and Gambardella (2004) explore the factors needed to become a successful software-producing country. They find countries showing "signs of vitality" when they:

- Have a higher share of educated population compared to their level of development
- Do not have a wide and diversified industrial base, which implies lower opportunity costs of these people to work in the software industry

[14] Gupta et al. (2005): "For example, in massive multiplayer games, if the event of triggering a player's rifle is delayed, that player may get killed and lose the battle". Also, video applications (video on demand, streaming) increase the demand for bandwidth and generate more traffic burst.

- Have domestic sources for the formation of software competencies (like Nokia in Finland)
- Believe in the importance of openness, which means favouring "free trade" and include the openness to multinational companies (MNCs).

A special case is the application of competition law on software companies (for example, on Microsoft; see Evans et al. (2005b)). Software patents are another special issue which are discussed in Sect. 9.3.2. The following subsections are related to standards, trade policy, and e-business initiatives.

9.2.2 Standards

Standards can be set:

1. By market power, often following a standardization war
2. By law or by regulation
3. Cooperatively. For example, the standard for Web browser commands is set by the World Wide Web Consortium which has over 400 members worldwide (W3C, see Sect. 8.2.3)

In a standard war, two or more holders of (intellectual) property rights try to enforce their standards. If the standard war is solved and a company holds a (de facto) standard, it has an incentive to restrict compatibility[15] to preserve market power (Church and Gandal (2005)). They also list strategies for a standard war in telecommunications. Adapted for IT services, the most important may be:

- Penetration pricing: Consumers are attracted by low prices, maybe below marginal cost. Or, as Church and Gandal (2005) put it: "subsidies to encourage adoption today are financed through higher prices in the future"
- Advertising and marketing: The model of two-sided markets in Sect. 6.3 emphasises the role of the initial marketing efforts
- Reducing the risks of a lock-in for consumers, so they are more willing to join
- Product preannouncements: Promised improvements may prevent customers from using a competing technology
- Complementary investments to increase the value of a standard.

Second, the combatants of a standard war may be supported by national governments directly or indirectly. For example, the government of Taiwan will invest USD 664 million to support the WiMax standard (see Sect. 2.3.1) in 2008 and 2009 (Hille (2007)). More examples can be found in Kaplan (2005).

Setting a standard by law or by regulation means that the development of a standard is driven by a governmental body. An example is the goal of the

[15] For the different forms of compatibility, especially of hardware-software systems, see Shy (2002).

Table 9.2 Types of standards used by companies in %; Source:EC (2006a)

Firm size	EDI	XML	Proprietary	Other
Micro (1–9 employees)	2	6	10	1
Small (10–49 employees)	4	5	13	2
Medium (50–249 employees)	10	10	24	2
Large (250+ employees)	29	27	31	7

European Union to establish a standard for e-invoicing, which should also be used internationally later and give "market certainty" to European companies (EEI (2007)).

The general equilibrium model in Chap. 5 shows that the support of standards is also a question of industrial organization and the capital:labour ratio of a country, even if the IT administration rights are set on firm level. Table 9.2 shows that the standards used for data exchange also depend on company size. The data are based on a questionnaire answered by 7,237 companies in 10 EU countries. If the share of international fragmentation (see Sect. 3.3.2) in a production process will increase further, automatic data exchange and the standard setting issue become more important also.

The general equilibrium model in Chap. 5 argues that labour-intensive countries should be more inclined to support open standards. According to Kaplan (2005), an: "open standard

- Cannot be controlled by any single person or entity
- Is evolved and managed in a transparent process open to all interested parties
- Is platform independent, vendor neutral and usable for multiple implementations
- Openly published, including availability of specifications and supporting material available royalty free or at minimal cost and
- Approved through due process by rough consensus among participants".

Third, the process of standard-setting can also be driven by privately-held companies. This development could lead to a closed (= non-open) standard or the participation in a standard-setting club. But UNCTAD (2006) points out that the attendance of companies is also a matter of firm size: "Although standard setting may be a formally open process, participation is costly and so the smaller player is priced out of the standards game and private considerations prevail over the public interest." On the other hand, UNCTAD (2006) emphasises the role of open standards especially for the evolvement of Web services and Software Orientated Architectures (see Sect. 2.3.2). IPTS (2001) hints at another problem of cooperative development of open standards referring to Web browsers: "The problem is that the W3C is not a software producer and therefore does not implement the standards that it produces. ... The W3C cannot interfere with commercial incentives in the use of its standards. The result is that computer browsers are becoming less inter-operable over time... Open Standards can lead to a widespread adoption of a new technology but it also illustrates the instability of open standards in the context of continuing high levels of technological opportunity." An alternative is the development of converters: Liu and Kemerer (2007) show that the availability of digital converters reduces the price premium of the leading flash card formats more than of the minority formats. Producers of the non-dominant standards can be better off with the provision of conversion technology, as this technology counterbalances the network effects of the market leader.

This last example also shows the tension between a non-open standard and the openness approach. A non-open standard is a mean to secure the exploitation of innovation rents, so that a trade-off between openness and innovation rents emerges here. But also a non-open standard could be contestable, as the strategy of the producers of the non-dominant standard shows.

9.2.3 GATS and IT Services

The GATS system is relevant for negotiating conditions of service trade (see Sect. 8.2.1). The Services Sectoral Classification Lists (WTO (1991)) determine the topics. Related to IT services, these are as follows (selected):

- Computer and Related Services
  - Consultancy services related to the installation of computer hardware
  - Software implementation services
  - Data processing services
  - Computer and Related Services

  For the EC proposal for a new classification of this sector, see Sect. 2.2.2. Other countries, Australia, Canada, Chile, Croatia, Hong Kong, Japan, Mexico, Norway, Peru, Turkey and the US among them, supported a slightly modified proposal in 2007 (WTO (2007a)).
- Telecommunications services
  - Packet-switched data transmission services
  - Circuit-switched data transmission services
  - Electronic mail
  - Voice mail
  - On-line information and data base retrieval
  - Electronic data interchange (EDI)
  - On-line information and/or data processing (incl. transaction processing)
- Audiovisual services
  - Motion picture and video tape production and distribution services
  - Radio and TV transmission services
- Other Business Services
  - (For example) Management consulting service

In a revised offer, from June 2005, the European Communities offers market access for the sector of computer services (for coverage of products, see Sect. 2.2.2) and National Treatment for all WTO member states for GATS Modes 1, 2, 3 and limited market access for GATS Mode 4 (WTO (2005a)). The offer is part of the so-called "Doha trade negotiations," which are still suspended (August 2008). Still unsolved is the classification issue of digital delivered goods (see Sect. 9.3.3).

9.2.4 National Initiatives and E-Government

There are a lot of national initiatives to promote IT and Internet services. A goal is to foster the impact of IT on productivity. UNCTAD (2006) reports that from 181 developing and transition countries, 80 had adopted a national ICT plan and 36 were in the process of developing one. Additionally, all developed countries have some kind of e-business, e-inclusion, e-payment e-government, e-commerce, e-health, e-library, etc. programmes.

In Europe, the current framework is called "i2020." It should support the completion of a "single European information space," strengthening innovation and investment in ICT research and achieving an inclusive European Information Society (EC (2005b)). A detailed overview of action of the EU commission and the national ICT policies can be found in EC (2007a).

The national programmes can be classified according to their focal points:

- "Infrastructure" of IT services, e.g.:
  - e-commerce laws such as digital signatures, intellectual property rights, etc.
  - e-payment initiatives, such as the introduction of the e-Dinar in Tunisia (UNCTAD (2006))
- Promoting skills (of pupils, workforce, the young, elders, etc.)
- Promoting IT-service applications in special sectors (e.g. health)
- e-government projects.

E-government projects are important to improve the service quality of government as well as to speed up the research and development of special IT services. According to the UN e-government survey (UN (2008)), 189 countries are online in 2008 (i.e. have a website). Recommendations for national e-government projects are given by UNCTAD, OECD, the European Commission and other institutions. For example, for any e-government services to be set up at a pan-European level, the EU Commission recommends consideration of the following general principles (EC (2004)):

- Accessibility: services should be publicly accessible without discrimination.
- Multilingualism.
- Security.
- Privacy: data protection.
- Use of open standards.

- Assess the benefits of open source software.
- "Multilateral solutions": a single solution should be developed once.

According to UN (2008), the next step is from e-governemnt services to "connected government." The concept of "connected government" should "advance the public good by engaging the creative efforts of all segments of society." On the other hand, OECD (2006a) points out that regulations for the re-use of public-sector information (e.g. geographical, meteorological, traffic data, etc.) is an ongoing development.

Finally, it should be mentioned that general business conditions are important to support the service sector also. For example, Rubalcaba-Bermejo and Gago-Saldana (2002) have analysed regional concentration with data from France, the UK, Austria, Belgium and Finland in 1997, finding that the business services economy is closely linked to urban density and other regulatory and instructional factors, but, at the same time, is linked also to the degree of exposure to international competition. So capital regions benefit from the highest presence of business services as they are the most global regions. Rubalcaba-Bermejo (2004) argues that many international flows in services require a local presence, and that local providers are the ones which can "glocalize" international services. This may be true for IT services as well and is in line with the predictions of the macroeconomic general equilibrium model of Chap. 5.

9.3 Digital Goods

9.3.1 Overview and Policy Challenges

The characteristics of digital goods are described in Sect. 2.2.3, and recent developments in Sect. 2.3.3. Economic theory related to digital goods is presented in Sect. 3.1.3 and Chap. 6. It is shown that the particular characteristics of digital goods lead to specific behaviour in macroeconomic models, also in the special case of two-sided markets (Sect. 6.3). Now emphasizing the role of digital goods for the so-called "knowledge economy," which describes the rising importance of culture, knowledge, and creative industries, OECD has identified six main areas of policy challenges related to digital content (OECD (2006a)):

- "Innovation and technology (e.g. enhancing R&D and innovation in content, networks, software and new technologies)
- Value chain and business model issues (e.g. developing a competitive, non-discriminatory framework environment)
- Enhancing the infrastructure (e.g. technology for digital content delivery, standards and interoperability)
- Business and regulatory environments..., protection of intellectual property rights and digital rights management...

- Governments as producers and users of content (e.g. commercial re-use of public sector information)
- Conceptualisation, classification and measurement issues".

This "policy menu" again shows that there are strong interdependencies between networks, IT services, and digital goods. In this thesis, most of the topics are covered in other sections (networks in Sect. 9.1, standards in Sect. 9.2.2, e-government in Sect. 9.2.4). Some selected examples of digital content initiatives of OECD countries are also listed in OECD (2006a) and ITU and UNCTAD (2007). The topics of intellectual property rights and digital rights management are covered in the next Sect. 9.3.2, followed by theory-driven sections about software patents and two-sided markets, and supplemented by Sect. 9.3.3 about trade policy issues.

9.3.2 Intellectual Property Rights (IPR)

9.3.2.1 IPR and Competition Policy

According to Sect. 2.2.3, it is a characteristic of digital goods that they are nonrival and can be copied at (almost) zero marginal costs. The protection of Intellectual Property Rights (IPR) is pivotal for the creators of digital goods to be able to realize economies of scale.

Generally, Farrell and Shapiro (2004) identify two polar viewpoints which represent different views about the sources of innovation and creativity: the incentives school, and the openness school.[16] Following the incentives school, IPR are justified as they encourage innovations. The owner of the right can exploit innovation rents, e.g. by selling licences. Ganslandt (2008) points out that there might exist tensions with competition law. For example, neither the US nor EC law grants antitrust immunity for companies exploiting their IPR. US and EC institutions accuse software company Microsoft of impeding competition although it is the owner of the IPR (see for US lawsuit, Evans et al. (2005b); for an EC lawsuit, Heinemann (2005)). On the other hand, IPR and competition law go hand in hand, as empirical studies show that countries with a high effectiveness of antitrust policy also grant sophisticated intellectual property protection (Ganslandt (2008)).

9.3.2.2 Copyright Protection and Digital Rights Management

Violations of copyrights can be found with all types of digital goods. A technical solution to protect the IPR is the introduction of a Digital Rights Management

[16] For example, the arguments of Lessig (1999) are typical for the reasoning of the openness school: "The best code (software) is both modular and open. Modularity ensures that better components could be substituted for worse. And from a competitive perspective, modularity permits greater competition in the development of improvements in a particular coding project".

system (DRM), which controls access to digital content files (see also Sect. 2.3.3). Examples are copy restriction tools on CDs or the registration procedure when installing new software. Bomsel and Geffroy (2006) interpret DRM as a private means of enforcing content excludibility, which participates in the "privatization" of copyright protection. As a result, DRM may lead to higher consumer welfare, when innovations occur, and variety and quality rise. On the other hand, Rayna and Striukova (2007) see a decline of welfare through DRM because of the higher bargaining power of suppliers, the inferior goods characteristics of DRM-protected goods (audio files cannot be lent), entry barriers for competitors, and the absence of a secondary market. They conclude that consumers are not willing to pay higher or equal prices to DRM-protected goods than unprotected ones.[17]

A starting point for economic policy are the transparency and interoperability of DRM systems (EC (2008)). Interoperability (e.g. to hear DRM-protected music on different devices) may limit the constraint of consumer utility.[18] While DRM emphasises the role of private provisions to protect IPRs, other approaches rely more on state-run institutions: For example, in France in November 2007, music and film producers, internet providers and the government agreed to establish a new institution which could block Internet access for users who download files illegally (EC (2008)). The US Digital Millennium Copyright Act (DMCA) establishes "a procedure by which a copyright owner can obtain a subpoena from a federal court ordering a service provider to disclose the identity of a subscriber who is allegedly engaging in infringing activities" (DMCA (1998)). Australia has introduced a scheme so that ISPs are required to remove access to the material or the link when they are notified by the copyright owners. This scheme is a result of changes following a Free Trade Agreement between the US and Australia (OECD (2006a)).

Internationally, the WIPO treaties (see also Sect. 8.2.3) are important to enforce IPRs. The WIPO treaties require "ratifying states to prevent circumvention of technological measures (such as DRM technologies) used to protect copyrighted works". The ratification of WIPO treaties into national law also include exceptions for special cases "involving non-profit libraries, archives, educational institutions, encryption research and others" (OECD (2006a)).

9.3.2.3 Software Patents in the Light of Theory

Another form of legal protection of an intellectual property right is a patent. In contrast to a copyright, a patent prescribes the disclosure of the invention and a patent

[17] This may have led to a change of strategy of music labels. In January 2008, various newspapers reported that EMI, Universal Music, Warner Music Group and Sony BMG now also support the distribution of DRM-free music.

[18] On the other hand, Park and Scotchmer (2005) argue that the implementation of a DRM system as an industry standard may lead to "collusion through technology".

can be granted also for ideas or new processes. Related to the topic of digitalization, the role of patents is debated primarily with regard to software.

Comparing the software with other markets, advocates (e.g. Huppertz (2004)), like opponents (e.g. Pilch (2004)) of software patents are both amazed that traditional patent-protected sectors like engineering have developed with sustainable growth rates for many years, while the software sector has succeeded to some extent without patent protection (especially without the possibility of enforcing attained patents).

The theory of incomplete contracts (which is introduced in Sect. 3.4) may solve this "secret". In engineering, an invention is mostly bound to a specific final product; thus, the outside option of the inventor amounts to zero in the worst case. Patents are necessary to stimulate an adequate investment volume. In contrast, software modules can be copied and used repeatedly. Therefore, the outside option of a software developer never amounts zero, which also stimulates the investment volumes without patents. From this point of view, the introduction of software patents is not necessary.

On the other hand, some countries may have incentives to support software patents. In the general equilibrium model about IT services (Chap. 5), a theoretical assignment is reached between the interoperability policy of a country and its basic economic data.

1. Capital-intensive industries show a higher inclination for the integration of foreign companies
2. Integrated companies prefer proprietary IT systems
3. Labour-rich countries have a stronger incentive to support open standards than capital-rich countries, although
4. Also from a theoretical point of view, integrated industries and trading (outsourcing) industries coexist.

This approach accounts for an economic policy which supports proprietary solutions, e.g. through patents, as well as open standards simultaneously. Consequently, different criteria are discussed (e.g. novelty) which could qualify a software to be patented (Bielig and Haase (2005)). Finally it should be mentioned, that software patents are not only a mean to secure the exploitation of innovation rents, but, in real-world business, are also used to constrain (potential) competitors.

9.3.3 Trade Policy

Classification issues seem to be boring compared to traditional economic theory of trade policy. But behind the struggle for the "right" classification, fundamental positions of countries or regions are recognizable.

9.3.3.1 Digitally-Delivered Goods: GATT or GATS?

WTO members agree that the majority of services that are delivered electronically (e.g. financial or professional services) are governed by GATS (see also Sect. 8.2.1). But they do not agree on whether digital products which have traditionally been traded on a physical carrier medium such as CDs and DVDs are goods (covered by GATT) or services, which are covered by GATS (Wunsch-Vincent (2005b)).

If the answer reads "GATS", further questions will arise: To which service subsector do electronically-delivered services belong? And: Which mode of service supply (see Sect. 8.2.1) applies? If GATS Mode 2, this would imply that the government of the supplier also decides on consumer-protection rules for consumers who are physically in another territory. For an overview of further issues which should be clarified by WTO's Work Programme on Electronic Commerce, see its 1999 Progress Report (WTO (1999)).

Presently, there might be no solution in sight regarding the classification issue of software and other digitally-delivered goods. The protocol of WTO's Sixth Dedicated Discussion on Electronic Commerce in November 2005 reads: "In this delegation's view,[19] the dedicated discussion's examination of the classification question, on which this group had spent considerable time over the past 5 years, had reached an impasse. The aim was to jump over that impasse and to focus on concrete issues that participants could grapple with in a more systematic fashion, and thus make progress on addressing something that was clearly relevant in today's trading world. Software was a $600 billion industry that was productivity-enhancing for all economies. Thus, all had an interest in its liberalized treatment. ... This delegation's goal was not to prejudice any Member's view on whether software was a good or a service, but rather to focus on the value of having a liberalized treatment for products that had traditionally crossed borders on a physical medium but were now increasingly being traded electronically. This was a discrete area where the WTO could, in a limited fashion, make concrete progress in setting a goal for a liberalized trade regime" (WTO (2005c)).

Nevertheless, different proposals for a solution of this classification issue have been made. For an overview of these proposals and a political assessment, see Sect. 9.3.3.

9.3.3.2 EC and US Positions in WTO Negotiations

In general, the EC and US declare they favour free trade. But the EC as well as the US "protect" specific sectors from trade liberalization. Meunier and Nicolaidis (2005) evaluate the European trade policy in the past as progressive, but agriculture, which is not in the scope of this paper, had "remained the glaring counter-example of liberalization". Wunsch-Vincent (2005a) identifies two regulatory contexts: on

[19] Delegation of United States

the one hand, the rules are aimed at progressive liberalization, deregulation and thus competition; on the other hand are the public-interest objectives like freedom of expression, pluralism, cultural and linguistic diversity. He argues that the requirements and rules for cultural diversity lead to the EU's refusal to put digital services under GATT rules but to favour GATS as an international system. In the case of the US, the water and energy sectors are examples which are not within the scope of the general liberal stance.

Regarding digital products, the positions of the EC and US happen not coincidentally, but are related to their respective legal or economic basics.

1. EC: A basic prinicple of the EC is to preserve cultural diversity and to protect and support the audiovisual sector (see Sect. 8.2.2 for the articles in the EC treaty). Therefore, it is argued by the EC that the audiovisual sector should be covered by the GATS and not by the GATT agreement (see Sect. 8.2.1 for the differences between them), so that exemptions can be scheduled. To avoid different WTO regimes for audiovisual products delivered electronically (via the Internet), broadcast or on a carrier medium, the EC suggests that GATS should be applied to all electronically-delivered products also.
2. US: In the 2003/2004 annual report of the US Trade Representative, the agenda after the ministerial conference in Cancun was lined out: "An aggressive agenda for market opening in services, ..., is being pursued in the negotiations. Since the United States is the world's leader in services for the twenty first century economy, and services account for 80% of US employment, our efforts in this area continue to be significant" (USTR (2003)). This approach includes digital products. It should be ensured that, when possible, the most liberal form of scheduling trade commitments is used (Wunsch-Vincent (2003)).

These differences in the legal or economic foundations became apparent in the WTO negotiations. Requests to other WTO members and the negotiating strategies are not published officially. Nevertheless, Krajewski (2005) and Freedman (2005) give an overview of the positions: Overall, the EC does not want to change the present status of the audiovisual sector. The EC identified telecommunications, water, business, financial services and transport as key targets of liberalization, but not the audiovisual sector. It makes no request in this area and it is not likely that it accept any essential requests from other WTO member. On the other hand, the US has requested binding commitments to liberalization in the audiovisual sector.

Consequentially, the EU has suggested considering all digitally-delivered goods as services. Wunsch-Vincent (2005a), p.172, interprets that: "this strive of the EC to afford the less liberal GATS treatment to digitally-delivered content products is meant to ensure that the EC can preserve its leeway for audiovisual policy intervention".

The US applies for a more liberal trade policy including the audiovisual sector. Therefore, the US argues for application of the GATT principle to digitally-delivered products. Also, the US suggested renewing the classification of audiovisual goods in the WTO negotiations with subsectors which are already listed in other sectors

Table 9.3 Differences in US vs. EC views on WTO digital content issues; Source: Wunsch-Vincent (2005a)

	US	EC
Moratorium	Make it permanent and have all digitally-delivered content products fall under this duty-free rule.	Make it permanent only if all digitally-delivered content products are considered services.
Valuation decision	Levy tariffs on the value of the carrier media and not on the value of the content.	No definite statement but the EC would only agree if all digitally delivered content products are considered services and thus outside of the ITA and the GATT.
Classification issue	Wants GATT-like treatment for digitally-delivered content products.	Wants classification of digitally delivered content products under the GATS.
In GATS	Classification of digitally-delivered content products under value-added telecom-services but not exclusively under audio visual services.	All content must be considered outside of telecommunication and computer services.
	Digitally-delivered software of any kind must be considered either under the ITA or-at the minimum-under computer services.	The EC invokes the audiovisual services-category for the classification of any content (except for business software).

such as telecommunications. This should help to enforce a more liberal policy in the audiovisual sector also. But the EC has rejected this proposal (Krajewski (2005)).

For an overview of the EC and US differences regarding trade of digitally-delivered content, see Table 9.3. As can been seen in line 1 also the political goals according a permanent moratorium on e-commerce (see Sect. 8.2.1) are different.

9.3.3.3 Prospects and Evaluation of the Classification Issue

Most WTO members do not apply tariffs, quotas, and other regulatory barriers to digital products. Although some members apply quotas or other measures in the audio-visual offline world (e.g. broadcasting quotas), this is not the case in Internet commerce. But the fear exists that some countries might change this in the future because the e-commerce moratorium is not binding (Wunsch-Vincent (2005b)).

This section does not answer the question whether free trade in all sectors may be beneficial for a modern society. To set a priority with respect to the two goals "to foster trade" and "to preserve cultural diversity" is first a normative question. Then, secondly, it can be analysed whether the instruments chosen are efficient.

Independently from the normative considerations, some economic criteria for a "good" trade policy within the WTO framework can be developed:[20]

[20] More general criteria for a trade policy framework are: market access, competitive market structure, compatibility, transparency, legal certainty, and dynamic efficiency (Krancke (2003)).

1. Trade policy should prevent distortions and has to be economically neutral (Bacchetta et al. (1998)).
2. To avoid uncertainty in companies and governments' policies, rules should cover new products or innovations automatically.
3. Trade policy goals should be within reach; that means they must respect the basic principles of contractual parties. Also, trade policy should take into account the different development levels of countries.
4. Trade is not a "zero-sum game". Thompson (2002) criticizes that "zero-sum game thinking" was typical for WTO member countries.

From item (1) it follows that digital content should be treated independently of the medium which carries the content (e.g. Internet or CD). From item (2), it follows that a negative list is superior to a positive list approach (see Sect. 8.2.1). From item (3), it follows that the basic principles of EC treaties (see Sect. 8.2.2) have to be taken into account, especially with respect to the cultural sector. Item (4) gives support to the economic arguments to imply the MFN principle also in favour of countries which are not members of the party.

The different suggestions for the classification issue are presented in Table 9.4. Compared to the present GATS system of general and specific commitments, the negative list approach is ambitious and forward-looking. Its advantages and disadvantages are explored in detail by Mattoo and Wunsch (2004), who analyse measures to pre-empt protectionism in service outsourcing. Related to the proposal in the last row, Krancke (2003) suggests differentiating e-commerce-related trading into three subsectors and assigning them to the trade regimes:

- E-business with physical goods and physical delivery, e.g. buying textiles on the Internet with delivery by a postal service be assigned to GATT.
- E-business with services produced digitally (electronically) , e.g. telediagnosis from a foreign hospital be assigned to GATS mode 2.
- E-business with digitalized products and electronic delivery, e.g. buying a music file be assigned to GATS.

Compared to the above-mentioned requirements for "good" WTO trade policy, each solution of Table 9.4 contradicts at least one goal. Technologically, neutrality is difficult to archive due to the ambivalent character of goods which can be delivered either online or on a carrier medium. Additionally, EC will insist on special treatment for the audio-visual sector. Therefore, a far-reaching solution with a new, hybrid product class also seems to be unlikely. On the other hand, the proposal to subdivide e-commerce trading and assign the subsectors to GATT and GATS does not basically solve the classification issue. The trading of textiles is assigned to GATT already (also when bought via the Internet), but the basic problem of different regimes for music delivered online and via CD would remain.

Leaving the analytic approach now and daring some speculation at the end of this section, I guess that a solution with repetitive prolongation of the e-commerce moratorium is the most probable solution from today's perspective. Alternatively, the moratorium may be changed into a more general framework agreement, which would not solve the classification issues but could clarify some aspects related

Table 9.4 Suggestions for future treatment of digitally-delivered goods

	Advantages	Disadvantages
Do nothing - prolongation of e-Commerce moratorium	Positions of US and EC are oppositional and no solution regarding digital delivery of audiovisual services insight. Prolongation of status quo would help to archive the other trade policy goals.	Moratorium is not legally binding; may pave the way for future restrictions harming global development of digital services.
GATT-Solution	Compatible with today WTO-Rules.	No technological neutrality compared to 'traditional' audio-visual broadcast services. Difficult classification problem. EC will not support this solution.
GATS-Solution	Compatible with today WTO-Rules.	No technological neutrality compared to delivery of audio-visual content on carrier medium. Long process to abolish trade restrictions through negotiations.
GATS-Solution+ negative list approach	New services will be covered automatically by commitments.	
New, hybrid product class for all goods and services which can be delivered digitally	Classification issue would have been solved.	Long and difficult negotiating process to establish new product class adjacent to GATT and GATS. Economic relevance (trading volume) of digitally delivered products does not justify own product class.
Further differentiation and assignment to GATT and GATS	Compatible with today WTO-Rules.	Overlapping of definitions, esp. in the audiovisual sector; no technological neutrality.

to national legislation such as non-discrimination, free access to information, and consumer access to e-commerce (Berkey (2002)).

9.3.4 Consequences of the Two-Sided Markets Approach

The two-sided market approaches presented in Sects. 3.2 and 6.3 may change the picture of content industries in the light of competition policy:

- Cartels have to coordinate on both sides. If only one side is fixed, competition on the other side would compete away the profits (Evans and Schmalensee (2005)).

- From this follows, that below marginal-cost pricing, which is common in two-sided markets, is not necessarily designed for foreclosing competition. Therefore, Evans and Schmalensee (2005) conclude, one should not presume that below cost-pricing was anticompetitive.
- Internationally horizontal mergers and acquisitions may enhance the consumers' welfare (see Sect. 6.3).
- Anti-competitive behaviour is difficult to prove as digital competitors may compete in quantities first, not in prices (see Sect. 3.2).

These arguments should have an influence on the assessment of whether a company misuses its market power. When one side of a market is subject to high non-transitory entry barriers, nevertheless the other side of the market may be competitive, which is also bound to an appropriate market definition. Last but not least policymakers should also take into account that Schumpeterian dynamics are a strong competition driving force even in a two-sided-market.

9.4 Convergence Issues

9.4.1 Overview

The item "convergence" describes a process as well as the final state of convergence. A final state is reached when a unified network transports all kinds of digital data, which may be voice, music, video, or any other kind of data. Next Generation Networks (NGNs, see also Sects. 2.3.1 and 9.1.4) is an oft-used description for this.

The process of convergence takes into account the different starting points. Table 9.5 shows which network technology has traditionally been used. Convergence in this context means that a service is distributed over a network which

Table 9.5 Technology and traditional regulatory principles of networks

Sector	Traditional network technology (see Sect. 2.2.1)	Most important regulatory principles (see Sect. 9.1)
Telecommunications	Circuit switched/one-to-one-connections	Universal service/price and access regulation of incumbents
TV broadcast	No switch necessary/one-to-many transmission via air directly to consumer	Must carry (see Sect. 8.1.4) in some countries: content quota and advertising quota
Data networks/internet	Packet-switched/any-to-any (= multiple) connections	Retail: access regulation (DSL)/wholesale: mostly unregulated markets

was formerly dedicated for another kind of service.[21] The networks become substitutes. Two examples are covered in this section: Voice over IP in Sect. 9.4.2 and "Broadcast over IP" in Sect. 9.4.3. Convergence raises specific regulatory concerns, especially how, if any, traditional regulatory principles of a service should be adopted. OECD (2006b) lists the problems when substitute services are subject to different regulatory requirements:

- Legitimate public policy objectives may be undermined, e.g. the financing of universal service obligation.
- Market outcomes will be distorted. Some firms may expand only because of regulatory advantages.
- "Regulatory arbitrage": Firms which face different regulatory regimes have an incentive to distort production decisions in order to be subject to a less-restrictive regime.

Furthermore, convergence itself may pave the way for anticompetitive behaviour. Nuechterlein and Weiser (2005) point out that cable companies are used to discriminating against TV content providers. It may well be feared that they may also want to pursue this strategy as access providers. Willig (2004) argues that bundling may be the "vehicle" for anticompetitive exclusion of rivals, especially if a bottleneck-holder refuses to grant access.

Finally, more mergers and acquisitions between companies can be expected because of convergence, especially if they serve different sides of a two-sided market (see also Sect. 6.3), or if a vertical "production chain" becomes apparent because of convergence between formerly-separated markets. An example might be the merger between internet provider AOL and media company Time Warner in 2000. Faulhaber (2005) mentions that in this case different remedies were applied by different institutions.[22]

Practice of competition and regulatory policy diverge between sectors and countries (see the example in Sect. 9.4.4). This contradicts the principle of "technological neutrality" (see Sect. 8.1.6). An integrated approach is presented in Sect. 10.2.

9.4.2 Voice Over IP

A general description of Voice over IP (VoIP) services is given in Sect. 2.3.4. The classification of VoIP providers is between:

[21] In literature, the term "convergence" is used non-uniformly. Lawson-Borders (2006) provides an overview and claims that convergence is an elusive term which is often ambiguous in its definition.

[22] The FTC (US Federal Trade Commission) has concentrated on the essential facilities and required open access to the cable, whereas the FCC (US Federal Communications Commission) has focused on the network externalities (see Sect. 3.1.1) and required (future) interoperation as a condition of the merger. Source: Faulhaber (2005).

1. Voice over Internet only providers
2. Gateway providers
3. Traditional PSTN and cable TV providers.

This was introduced there emphasising the ambivalent character of VoIP with respect to its assignment to digital goods or IT services. The ambivalent character of VoIP surfaces again when regulatory questions come into concern, especially whether VoIP services should be regarded as a telecommunication or an Internet service. The different regulatory regimes (see Table 9.5) meet here.

On one side, PSTN regulation may have an impact on VoIP. If classified as a telecommunications service, VoIP providers have to follow the principles of the PSTN market. The most important are:

- Universal service obligation
- Obligation to interconnect/terminate, which comes along with (regulated) access charges
- Carrier selection and carrier pre-selection obligations
- A special numbering system, e.g.:
  - Unequivocal emergency numbers
  - Local geographic numbers
  - Number portability
- Security obligations, e.g. storing connectivity data of users.

If classified as pure Internet services, VoIP providers could remain unregulated. The treatment of VoIP by OECD countries is not uniform and the market definition issue is still open.[23] As a result, in most countries VoIP providers have to follow some of the PSTN obligations listed above. Some issues are more complex with VoIP, e.g. the routing of a "nomadic emergency call". It could be routed to the emergency centre next to the VoIP provider, next to the caller's home, or next to the caller's current location provided it can be identified (ERG (2007a)).

The different approaches – to issue a separate licence or to look at VoIP as a traditional local voice service – are "well expressed in the numbering policies" (Fowler (2008)). The burden of using local geographic numbers would show the regulator's orientation towards "facility-based competition" or "maximum competition in the local market". But a mixed strategy is also possible, e.g. in Germany, VoIP providers can allocate local geographic numbers as well as numbers beginning with initial numbers 032 (Bennekoh and Holznagel (2005)).

The other side of the coin is the influence of VoIP on the PSTN regulatory regime as VoIP services are based on packet switching and the Internet Protocol (IP).[24] More concretely, VoIP technology impacts PSTN regulation because of the following.

---

[23] OECD (2007e) avoids a predefinition: "Such market definition issues may be best dealt on a case-by-case and empirical basis".

[24] For the general differences between Internet and PSTN impacting regulatory models, see Sect. 9.1.2.

- Price structure and interconnection regimes will change. Interconnectivity of data networks is often granted at zero costs (peering; see Sect. 9.1.2). Alternatively, a capacity-based interconnection pricing system may evolve (Kennet and Ralph (2007)).
- No termination fees are usually paid for VoIP calls to (other) Internet users. But VoIP users have to pay for call termination in a PSTN network. The negotiated fee paid by the VoIP provider for accessing the PSTN network also influences the retail prices as shown by Bijl and Peitz (2006). Furthermore, OECD (2006b) mentions that VoIP could also threaten government revenues because in many non-OECD countries the revenues of state-owned, monopolistic telecommunications companies are a key source of governmental budgets and foreign currencies. Many non-OECD countries are net importers of international phone calls which can now be substituted by VoIP.

If a PSTN incumbent enters the VoIP market, the danger that it will squeeze out new entrants (e.g. by offering predatory bundling (OECD (2006f)), or by technological foreclosure (incumbents influence protocols and numbering systems (Sadowski and Straathof (2005)))) is obvious when the regulatory regime does not countervail these. So ERG (2007c) emphasises the importance of bitstream access[25] for VoIP. Generally, economic criteria to decide upon imposing regulation could be (Monopolkommission (2006), OECD (2006e), and Sadowski and Straathof (2005)) the following:

- The conditions of market entry, e.g. the existence of entry barriers.
- The degree of substitution between VoIP and traditional voice services.
- The degree of platform competition in the retail market (as Internet services can be decoupled from the underlying network), e.g. the degree of convergence between fixed and mobile services.[26] Also, the degree of economic integration between VoIP, access (e.g. DSL, based on LLU or bitstream access) and PSTN provider.
- The degree of competition in the wholesale broadband Internet access market (the conditions of interconnection).
- The prevailing origination and transport protocols.

OECD (2006e) particularly recommends excluding VoIP from particular ex ante regulation for the retail voice market, while ensuring that broadband Internet users have unrestricted access to Internet applications and services including VoIP. More generally, the question about VoIP regulation is a question about the importance of the regulatory principle of "technology neutrality" (see Sect. 8.1.6). If this principle is imposed, VoIP and PSTN should have a common regulatory regime. This could vary from strict PSTNlike regulation to relatively free Internet regulation.

---

[25] See Sect. 9.1.2 for a definition.

[26] In some countries, dual-mode handsets use the Wi-Fi protocol (see Sect. 2.3.1) to acess VoIP through DSL connections in the home environment. See OECD (2007b)

For a general approach of competition policy in the light of convergence, see Sect. 10.2.

9.4.3 Broadcast Over IP

Traditionally, broadcasts have been delivered terrestrially. Newer delivery technologies use cable or satellite.[27] A common characteristic is "one-to-many" communication, so that the broadcast is sent once and received by many, without damaging the signal quality. In technical terms, there is no rivalry in consumption.

Delivery over IP networks may change this pattern, e.g. with unicast, (small but) positive marginal costs related to the number of viewers arise. The following distinctions are used (OECD (2007h), OECD (2004a), and Wegner and Bachmeier (2000)):

1. Quality and bandwidth
   - *Web TV or streaming*: Videos (or programmes) which are embedded in Internet pages, usually with different qualities (speed) for smallband users (e.g. 100 Kbit/s) up to broadband users (300 or 750 Kbit/s).
   - *IPTV*: TV programmes which are transferred via internet protocol. This technology supplements distribution via air, satellite or cable, as it uses a DSL Internet access, for example.
2. Router technology
   - *Multicast* means to deliver the broadcast signal once and then to replicate it from router to router. This technique has not yet been implemented as a standard.
   - *Unicast*: Standard technology today. Each user has a separate connection to the video server. The necessary bandwidth is proportional to the number of parallel viewers.
   - *P2P: peer-to-peer*: A client, e.g. consumer's PC, acts as a server for another client. Security and speed problems limit the practicability of this technique.

   Unicast needs high investments in server technology. Multicast and P2P may reduce the total bandwidth necessary.
3. Program chronology
   - *Linear:* Such as TV programmes, behave traditionally, with specific content at a point of time. OECD (2007h): "Audiences initially did not really choose broadcast content, it just arrived in their homes".
   - *Nonlinear:* Content can be chosen on demand.

[27] Satellite markets are not considered further in this thesis. Because of rising capacity, digitalization and the strict separation of network and content, Wagner (2004) calls them "boring" from a regulatory perspective.

Table 9.6 Impact of communication and content regulation

Telecommunications	Internet	Broadcasting
One-to-one communication	Any-to-any communication	One-to-many communication
Single telephone call has no public impact	Increasing impact on the public	Impact on the public. High relevance for democracy and society
Content of the telephone call is not regulated	Public content must not violate public law. No ex ante regulation	Public service broadcasters and content regulation in many countries

Now with convergence, the different policy basics of telecommunications, the Internet and broadcasting coincide (see Table 9.6). The presumed "power of pictures" and scare frequencies for terrestrial delivery let governments regulate broadcast content (see also Sect. 8.1.4). The pivotal question now is which regulatory principles from traditional TV markets should be applied to the distribution of audiovisual content over IP. These are the "must-carry rules" and "content quotas" predominately.

Noam (2006) argues that the Internet will be the major delivery platform for broadcasts in the future. Therefore, telecom regulation would become the foundation of media regulation. A more complex approach is pursued by the European Commission: the "Audiovisual Media Services Directive" differentiates between "television broadcasting" (linear) and "on-demand audiovisual media" (non-linear) services. For example, TV broadcasters should be granted more rights to access "events of high interest to the public" even if another broadcaster has an exclusive right. This right is not granted to on-demand services producers. On the other hand, the prohibition of tobacco advertisements should come into law for both groups of producers (EC (2007b)).

9.4.4 International Policy Aspects

Traditionally, telecommunications and broadcast licences were provided by national (or regional) agencies for national (or regional) markets. This type of market separation is neither feasible for TV signal distribution via satellite[28] nor with international content distribution via internet. Although technologies exist to prevent content being viewed by unauthorized users, technologies also exist to circumvent these restrictions (see also Sects. 9.1.5 and 9.3.2). As OECD (2007h) points out: "Internet has blurred the line between 'national' and 'international' content.... A proxy

[28] The EU has reacted to this with the "Television without Frontiers Directive". Its "country of origin principle" states that "audiovisual media service providers are only subject to their own Member State's jurisdiction". For more information, see http://ec.europa.eu/avpolicy/reg/avms/index_en.htm

server on a distant island is enough to web cast any content to anywhere". This affects audiovisual content and telecommunications services as well, since a third-country VoIP provider can be used to establish a connection between two users in two countries.

Internationalization of media content started early. For example, with satellite technology, global broadcast TV networks became available in the 1980s. Content providers are BBC World, Deutsche Welle, CNN, TV5, for example.

Two-sided market dynamics drive international mergers and acquisitions, as the model in Sect. 6.3 shows. Additionally, convergence leads to new opportunities in the search for economies of scope and scale, reducing transaction costs or the stimulation of service or content innovations (OECD (2004a)). International price structures are affected, e.g. VoIP could exert pressure on international termination fees of telecommunications providers (OECD (2006b)).

Summarising, convergence puts questions of international competition and regulation policy on the agenda. The ambivalent relationship between preventing the misuse of market power and fostering innovations is now an international topic too. As questions of intellectual property rights and ownership, price structure and market power, technological bottlenecks and content distribution are interrelated and competition authorities have been organized within the national boundaries, new regulatory challenges occur. Even convergence of markets could lead to greater divergence in the regulatory approaches. Vanberg (2007) compares US and EU reactions to convergence with regard to regulation of the telecommunications sector: In the US, regulations were "curtailed dramatically" since 2003, because it was recognized that infrastructure platform competition has increased significantly. In Europe, instead, the scope of the regulatory framework was extended to electronic communication.[29]

However, the two-tier competition policy framework of the European Union, which takes into account supranational and national implications as well, might be a noteworthy model for a consistent international approach. An important feature of EU regulation is the common set of product classifications and market definitions.[30]

[29] However, the list of relevant markets which may become subject to ex ante regulation have been reduced in December 2007 (see Appendix 12.8). For the framework "Directive for electronic communications networks and services" see Sect. 8.2.2 also

[30] The importance of a common classification system also shows the GATS/GATT debate about digital goods; see Sect. 9.3.3.

Chapter 10
Integrating Theory and Policy

10.1 Conclusions from Parts I–III

There is a vicious cycle of dynamic interdependencies between macroeconomic variables, firm behaviour, and the policy set-up. Policy influences business strategies; market outcomes influence macroeconomic figures and policy. Regarding networks, IT services and digital goods, this cycle is especially driven by new technological developments – and specific skills. Pilat and Devlin (2004) emphasize the self-enforcing effect: having a strong ICT sector may help generate skills needed to benefit from ICT use.[1]

Having analysed policy issues in Chaps. 8 and 9, the theoretical "predictions" from Fig. 7.5 can be considered plausible: standardization (Sect. 9.2.2) and interconnection (Sect. 9.1.2) regimes are crucial areas of political work, support for domestic companies in competing for global leadership positions can be presumed (Sect. 9.2.4). The promotion of IT services by supporting non-IT sectors is an implicitly-accepted strategy (Sect. 9.2.4). As the international diffusion of networks, IT services and digital goods is going on, political institutions should feel the urge to harmonize their approaches (see Sects. 8.1.5 and 9.4.4 for examples).

Overall, theory and policy chapters emphasize the overwhelming importance of technology, here appearing as the concept of digitalization (Sect. 2.1): Networks favour fragmentation of trade (Chap. 4), and the organization of supply chains is supported by IT services. The IT administration right is equivalent to the right to organize the company's processes (Sect. 3.4). Huge economies of scale, which are a result of the characteristics of digital goods (Sect. 2.2.3) lead to globalization strategies (Sects. 6.2.2 and 6.3). Macroeconomic variables are changed by technology (e.g. the total factor productivity figures, see Sect. 2.4), and they determine which technological approach is favoured by countries with different labour/capital endowments (Sect. 5.3). Finally, digitalization changes the shape of the sectors themselves:

[1] But it should also be taken into account that policy instruments always have different transmission channels. For example, Ward (2007) found evidence that the number of music stores fell when high schools received subsidies for Internet connections. He concludes that, "this intervention in education policy could have contributed greatly to the decline in the music industry."

M. Vogelsang, *Digitalization in Open Economies*, Contributions to Economics,
DOI 10.1007/978-3-7908-2392-9_10,

convergence (Sect. 9.1.4) is a topic of technology which leads to a reshaping of sector-specific policy instruments.

The theory and policy chapters as well disclose the dichotomous characteristics of networks, IT services and digital goods. Either they can be considered as a general-purpose technology (Sect. 2.1.2), then the spread of networks, services, digital goods and general political objectives are strongly related (Chap. 8); or they can be viewed as economic sectors with their own rules of the game (which is emphasised in Chap. 9). Both perspectives do not contradict and coexist.

Digitalization favours internationalization and vice versa (Sect. 7). Generally, economic policy does not impede this international diffusion of networks, IT services and digital goods. But some exceptions exist, for example, when cultural beliefs are touched (Sect. 8.1.4). The international WTO trading rules (Sect. 8.2.1) allow unrestricted trade in network goods (hardware) and support e-commerce. But IT and telecommunications services are subject to the more complicated GATS system. The recent debate about the classification of digital goods has only a minor impact on trading volumes. Furthermore, governments try to support their national IT and media industries, also in standardization races and to gain the rents of innovation (Sect. 9.2.2). But networks, services and digital goods have been subject to different regulatory regimes historically (Table 9.5). As services can be delivered via competing networks now (convergence, e.g. VoIP or broadcast over IP, see Sect. 9.4), it could be argued that the sector-specific political concepts should converge also. So three important topics, which have to be addressed by economic policy related to digitalization, are: How to solve the trade-off between openness and the necessary exploitation of innovation rents? Should a national or international regulatory approach dominate? How to change the regulatory rule set in the face of convergence? Especially the last question comes into the focus now when the new layer-orientated policy proposal is discussed.

10.2 Proposal for an Integrated Policy Perspective

As argued, convergence necessitates a reshaping of policy instruments: telecommunications, data networks and broadcast policies historically rely on different fundamentals which clash now (see Tables 9.5 and 9.6). The proposal for a new approach is based on three governance principles, which have been introduced in Sect. 8.1.2.

- Layer-consistent governance
- Technology-neutrality of regulation
- Non-discrimination/network-neutrality.

This section now explores the consequences of a symmetrical regulation of technological layers, rather than along different industries as is usual today. Noam (2006) calls this "techie school of thought". Figure 10.1 shows the structure of an accordant "digital policy workhouse".

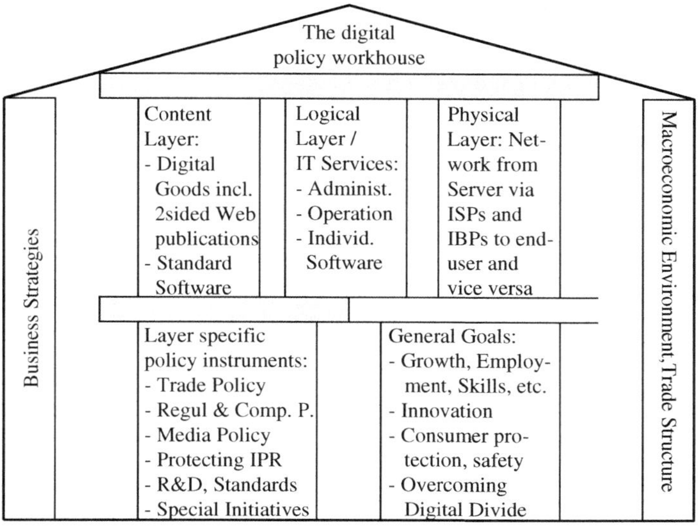

Fig. 10.1 The digital policy workhouse

In Fig. 10.1, all existing sectors are assigned to three layers.[2] It gets on the separation of networks and services which is a consequence of IP technology (see Sect. 9.1.5 also). Using this model to build the economic policy framework, there are no convergence-induced inconsistencies anymore.

The idea of structuring policy along layers is related to the proposal of OECD (2004a), that industry-specific regulation should be along "activity lines". According to OECD, a broadcast provider should have to acquire two licenses: for the carriage of the TV signal, and for the dissemination of content. Also, access requirements (such as local loop unbundling) for cable providers have to be introduced.[3]

To describe the proposal and explore its consequences more detailed, the following questions are taken into consideration now:

1. Does this framework imply technology-neutrality of regulation?
2. Does this framework lead to non-discrimination?
3. Can sector-specific policy be carried over smoothly?
   (a) How to finance universal services?
   (b) How to consider cultural aspects?

[2] Other authors propose differentiation between four or more layers. For example, Werbach (2002) suggests structuring policy along "content", "applications or services", "logical" and "physical" layers. But the more the layers that are introduced, the more difficult it will be to solve questions of classifying specific services.

[3] As introduced in Korea, the Netherlands and Canada (Wu (2004)).

4. Can the framework handle misuse of market power, e.g. cross-subsidization,[4] which may lead to a squeeze out of competitors?
5. Can this framework be applied internationally?
6. Does it solve the trade-off between openness and the exploitation of innovation rents?

Ad (1): Yes, assumed that digital goods, services and networks can be supplied independently of each other. The Internet Protocol allows the separation of services and content from networks. This separation is a question of standardized and open interfaces (see also the characteristics of an open standard mentioned in Sect. 9.2.2). Then, the layer principle can be applied. For example, all networks are subject to the same regulatory regimes, without influencing the other layers. Layer crossings in regulatory practice are minimized.

Ad (2): In parts. The support of open standards (Ad 1) can also be interpreted as a discrimination (against companies which try to establish closed standards). Furthermore, as shown in Chap. 5, labour-rich countries have a greater incentive to support open standards than capital-rich countries. Second, network-neutrality might be circumvented by new technological requirements (see Sect. 9.1.5). On the other hand, the recent framework, which distinguishes between broadcast, telecommunications and the Internet sectors, means more severe discrimination caused by regulation. Waverman (2006) emphasizes that non-discrimination should not be a goal of ex ante regulation, because most discrimination was actually differentiation and innovation.[5] Instead, competition law on a case-by-case basis should be relied upon more.

Ad (3): Mindel and Sicker (2006) and Frieden (2003) argue that the EU framework (see Sect. 8.2.2) is a step from the old "silo" model (regulation for every sector) towards the layered model, as services and conduits are separated in the framework. But in practice, a regulation (regime) is path-dependent.[6] Furthermore, the ground floor and first floor in Fig. 10.1 are not independent: expected policy measures influence actual behaviour of companies. Also, this layer-orientated framework cannot answer all questions of economic policy. For examples, it remains to discuss the role of innovation and innovative rents, definition of relevant markets, role of contestable markets, etc. But to the opinion of the author, the enforcement of long-term regulation objectives is a better strategy than to risk layer-specific distortions. Questions 3(a) and 3(b) shed light on the two policy instruments mentioned in the left column of Fig. 10.1 pars pro toto. A more complete evaluation

---

[4] Horizontal or vertical market power, for example: IBP and incumbent of telecommunications sector (Deutsche Telekom AG) offers Pay-TV and video on demand; a cable company (Time Warner) merges with ISP and portal (website) provider (AOL); a search engine (Google) wants to enter the mobile phone business.

[5] Waverman (2006) uses the example of an exclusive Premier League sports programme: this is discriminatory, but not anti-competitive.

[6] For example, it might be expected that cable operators will protest against access requirements even if those are imposed on telecommunications operators already.

of consequences of the layered model for single instruments is provided by Whitt (2004).

Ad (3a): Specific (financing) obligations could be imposed within the layered model also. For example, Whitt (2004) recommends that the responsibility for a universal service fund should reside at the "physical" (in the context of Fig. 10.1: "network") layer.

Ad (3b): The target of media policy (culture, youth protection, etc., see Sect. 8.1.4) would be the content layer. The layered model also implies that different content regulation across media types is disputable. Furthermore, international coordination would become a more important topic of media policy.

Ad (4): If there is a misuse of market power, the case should remain a matter of competition policy. The layer-consistent governance principle is applied ex ante, not ex post (see Table 8.2). But it will also dampen misuse of market power, as "new" and "old" competitors of a product are subject to the same framework, and vertical market power will be limited by the separation between services, goods, and networks. The more severe problem is the exploitation of existing customer bases, especially in two-sided markets (Sect. 6.3). This could lead to a transfer and a misuse of market power in new segments, which could not be prevented by an industry-specific regulatory regime nor layer-orientated governance. But the implementation of the non-discrimination principles (freedom to access content, to run applications, to attach devices, to obtain service plan information; see Sect. 8.1.2) supports sovereignty of customers, which counterbalances a company's market power.

As industry-specific regulation may suffer a loss of acceptance within the layered model, competition policy remains a crucial tool. Kariyawasam (2007) shows that the fundamental concepts of today's competition policy can be retained. He argues that the concept of significant market power (SMP) does not need a new definition, but (only) a new way of interpretation, and recommends applying the SSNIP test (see Sect. 8.1.6) at each layer.

Ad (5): Yes, the framework could be applied internationally. This is an important feature, also because convergence drives the internationalization of networks, IT services and digitals goods (Sect. 9.4.4). Furthermore, because digital delivery and digital content production show enormous positive scale and network effects, a plurilateral framework (such as the WTO) might be considered as the "natural" framework for networks, IT services and digital goods. On the other hand, Piazolo (2001) argues that the Internet is decentralized, and, therefore, decentralized regulation would be suitable – even if it will be necessary to increase coordination at multi- or bilateral levels.

To solve the trade-off between the national and international regulatory approach, Europe's subsidiarity approach could be a solution: Its two-tier model (Sect. 8.2.2) implies that general guidelines are introduced supranationally, and specification and implementation are done regionally (nationally). In telecommunications, a two-tier model was accomplished by ITU (International Telecommunication Union, Geneva, formerly International Telegraph Union) in 1865. But as ITU is primarily occupied with technical harmonization, convergence now gives reason to think about a multilateral two-tier model for all policy instruments mentioned in the left area of the ground floor in Fig. 10.1. Then, digitalization would have led not only to an internationalization of business, but also to an internationalization of an appropriate accompanying political framework.

Ad (6): In parts. The proposal requires the openness of interfaces of networks, digital goods, IT services as far as it is necessary to separate services and content from networks (Ad 1). Furthermore, this separation and the non-discrimination principle imply network-neutrality. Network neutrality prevents the discrimination of specific content from specific sources by the network provider, but allows the differentiation of prices according to customer's bandwidth. Other topics such as the intellectual property rights of IT services and digital goods are not explicitly addressed by this proposal but remain subject to flexibility, e.g. layer specific policy goals and instruments as mentioned in Fig. 10.1.

10.3 Policy Challenges

The layer-approach introduced is driven by convergence, a technological development, and based on the principles of technological neutrality, non-discrimination and layer-consistent governance. However, although a new conceptual approach to policy would be meaningful in the age of digitalization, some political imponderabilities remain. As described in the previous chapters, trade-offs exist between innovation rents and the openness of standards or the neutrality of networks respectively, as well between international political coordination and national interests. Technological dynamics, international political spillovers, company and consumer interests, as well as country-specific strategic aspects always drive to some extent the economic and political debate.

But, above all, policy development might be too slow to account for technological developments. Therefore, it is necessary that politics rely more on general political principles, but not on industry approaches predominately. Based on principles as mentioned before, policy reviews could occur more frequently, without having to fear erratic policy moves.

International spillover occurs in two ways: by the international diffusion of networks, IT services and digital goods; and by the accompanying policy instruments. This is a strong argument for international bodies such as the WTO or the ITU to obtain the right of regulation. On the other hand, national governments are not keen on transferring political rights without expecting a net benefit. This approach could be replaced by a club approach, where several leading countries start with a new regulatory concept. Then other countries might be more inclined to rethink policy to make possible a consistent global policy concept. This aspect is strongly related to the new political economy, which stresses the self-interest of politicians and administrations. According to this view regulatory bodies are ever-expanding.

Finally, networks, IT services, and digital goods can be used to achieve military or political goals. This is a strategic view of digitalization which is not in line with the economic perspective. Upcoming developments such as the possible use of artificially intelligent robots in armed hostilities may aggravate this tension. A two-tier approach as proposed above could be the best instrument to encourage international coordination and the willingness to compromise.

Chapter 11
Summary

This study explores the interdependencies between digitalization, economic policy, and macroeconomic variables of open economies. Digitalization describes the dismantling of information into strings of 0 and 1. To study the phenomenon of digitalization with the instruments of economics, it is split into three components: networks, IT services, and digital goods. This separation corresponds with the layer approach which is based on technological characteristics of data package transport and use via the Internet protocol (IP).

Existing economic literature is extensive and detailed about the regulation of telecommunication networks, but little research has been done on IT services and digital goods. Therefore, a focus of this study is on the development of elements for theory, as well as the description of existing theory elements from related areas. The model of IT administration rights and the description of the dynamics of two-sided markets are newly created. Existing model frameworks, e.g. from trade theory, are then supplemented with these new components to be able to formally describe the cross-effects between digitalization and macroeconomic variables of a country.

The selected frameworks are general equilibrium models, since they are the best choice to explore the interdependencies between different markets under optimizing behaviour in a macroeconomic set-up. The model in Chap. 4 takes into account the global market, the model of Chap. 5 is a two-country model, and that in Chap. 6 is first developed for a closed economy, to be opened for foreign suppliers of digital goods in a second step. From a theoretical point of view, the challenge is the integration of networks, IT services, and digital goods in the macroeconomic frameworks. Regarding IT services, the new theory component of IT administration rights builds a bridge between the sector-specific and the macroeconomic variables. Regarding digital goods, this task is done by the profit optimization strategy of a two-sided platform supplier. The assumption of communication requirements when intermediates are traded between intermediate and final goods producers fills the gap in the network model. The differences between the models are analysed in Chap. 7 in detail.

The theoretical analysis shows how macroeconomic and sector-specific variables affect each other. In the model of Chap. 4, digitalization leads to an increase in the volume of trade in goods, and the number of enterprises which offer products only for the domestic market is reduced. From this, it follows that the conditions of global

M. Vogelsang, *Digitalization in Open Economies*, Contributions to Economics,
DOI 10.1007/978-3-7908-2392-9_11,

interconnection of networks become important from the perspective of general economic policy. The model of Chap. 5 shows that the relative endowment of a country with capital and labour affects the trade of IT services. Companies with headquarter in relatively labour-rich countries prefer outsourcing solutions, while enterprises headquartered in relatively capital-rich countries are more inclined to organize their IT service departments in-house (offshoring). In digital goods markets, especially with two-sided markets, the market entry strategy of foreign competitors is dependent on the size of a country. Large markets are first exposed to foreign competition (Chap. 6).

The main result of the theory chapters states that links exist between the macroeconomic set-up and the pattern of digitalization. Therefore, macroeconomic policy influences the sectors of networks, IT services, and digital goods, as well as sector-specific economic policy influencing macroeconomic variables. One may argue that this is also the case with traditional sectors, e.g. steel production or bank services, but digitalization deserves closer attention. First, the flow of digital information, e.g. an eMail or a software, can be delivered worldwide at nearly no marginal costs in a millisecond. Second, digital strings can be reproduced at no marginal costs. As the international availability of the Internet fosters international trade, as shown in Chap. 4, one may expect that production processes harmonize. But opposite effects are at work also: trade of IT services depends on the capital-labour ratio of a country, as shown in Chap. 5, and digital goods imply strong first-mover advantages, which is also true with two-sided markets, as shown in Chap. 6. From this point of view, digitalization has an ambivalent character: the international spread of networks fosters trade and the international harmonization of production processes, but the international trade of IT services and digital goods goes on with existing patterns of specialization.

Generally, digitalization leads to internationalization, and vice versa. Although a lot of anecdotal evidence for the results of the theoretical models presented can be found, an econometric confirmation of the results of Chaps. 4 and 5 is not possible due to the non-availability of data or time series which are too short. Regarding networks, an empirical analysis is presented: a relationship between telephone minutes and exports, both total and per GDP, can not be disproved from the data. However, the relationship between exports and Internet bandwidth or number of Internet users respectively is less clear. A reason could be that for the negotiations and deployment of commercial transactions, e.g. by email, large bandwidth is not necessary.

Next, the implications of the presented models for economic policy are compared to real-world politics. As shown the design of international economic policy for networks, IT services, and digital goods also influences the macroeconomic performance of a country. Therefore it is not surprising that national and regional institutions as well as international bodies try to influence economic policy in the field of digitalization. For example, standardization and interconnection regimes are crucial areas of political work. Also, support for domestic companies in competing for global leadership positions can be presumed. In the following chapters, both, general goals of economic policy and sector-specific approaches are examined.

A trade-off becomes apparent in various policy chapters: the trade-off between open standards, interfaces and networks on the one hand, and the legitimate country-, sector- or firm-specific interests to realize innovation rents or just to favour proprietary solutions on the other. Related are regulatory questions, which also affect the strategic dimension of the international distribution of political and economic power: For example, ICANN, the institution which is responsible for the assignment of IP numbers, acts under US dominance. Since IP numbers are the pivotal element of the global Internet address system, this attracts criticism from other countries.

Overall, also sector-specific policy approaches can be related to more general interests. The fundamental legitimations for an active economic policy are presented in Chap. 8. These encompass, for example, the protection of the young, cultural promotion (media policy), the protection of abusive use of market power (competition policy), and trade policy. Within the trade policy, it is shown how national interests affect the development of the international WTO framework, even when questions of definitions have to be solved. Overcoming the digital divide, which can be differentiated between the digital divide within societies or between countries, is a goal of social and development policy.

In Chap. 9, the sector-specific policy approaches for networks, IT services, and digital goods are presented. In Sect. 9.1, the markets for network access and interconnection are in the focus. Access describes the connection of end-users and content providers; interconnection is the data exchange between Internet Service Providers and Internet Backbone Providers. Following the literature on telecommunication networks, there are strong arguments for market regulation, especially when a last-mile monopoly exists. But a different network topology of digital networks (IP networks) compared to the traditional telecommunication networks leads to stronger competition for Internet providers, also for former telecommunications incumbents. However, regarding IP networks also, the regulatory regimes for wholesale and retail markets are interdependent. A specific regulatory approach for digital networks is the principle of "network neutrality", which follows from the principle of non-discrimination. As shown in the theory part the costs of communication also influence the international volume of goods trade; therefore, the set of rules determining access and interconnection are crucial not only for IT services and digital goods, but for all sectors of an economy.

The model about IT services has shown that according to the relative endowment of labour and capital, different types of IT services trade are preferred. From this, it follows that countries will try to support domestic companies. Section 9.2 confirms this and analyses the different approaches of national IT services initiatives, e-government strategies, and the active deployment within standard races following the goal of introducing dominant standards. Furthermore, IT services are subject to the different, country-specific positions within the GATS negotiations of the WTO framework.

The theoretical approach presented about digital goods rationalizes the importance of market leadership for providers of digital content, especially of two-sided markets. Section 9.3 describes the possibility of defending intellectual property rights by the introduction of digital rights management (DRM) systems and software patents. The set of rules regarding the international trade of digital goods also has impacts on other important topics: network neutrality again will be affected if Internet Services Providers are forced by law to control content for property rights violations. The trade policy question of whether digital goods should be ruled by the GATS or the GATT regime affects the openness of relevant markets. The European Union prefers the GATS regime, so that existing national rules, for example, regarding cultural diversification in TV programmes, can be maintained. So country-specific positions regarding the mentioned trade-off between openness and the exploitation of innovation rents do not depend on technological considerations alone, but also depend on the general economic policy considerations, the capital-labour endowment ratio, and cultural and strategic aspects. They determine which degree and direction of openness of a market or a technology is preferred. In part it is a question of strategic trade policy, but it should be taken into consideration also that some kind of openness could also be wanted at country or firm level, for example, if revenue streams can be generated by open-source software projects (Sect. 2.2.4).

Now digitalization leads to a second, completely new regulatory challenge: convergence. TV, telephone and Internet services are no longer restricted to specific types of networks, but due to digitalization, networks and services can be separated. Examples are Voice over IP (VoIP), or the download of videos via telecommunication cables. Convergence exerts pressure on rethinking the traditional sector-specific regulatory approaches. Convergence also demands international cooperation, so that, for example, youth protection could also work when videos are downloaded from stations or servers in foreign countries.

So three questions have to be addressed by economic policy related to digitalization: How to balance the positions regarding the trade-off between openness and proprietary solutions, or the necessary exploitation of innovation rents respectively? Should a national or international regulatory approach dominate? How to change the regulatory rule-set in the face of convergence?

In Chap. 10, theory and policy findings are consolidated and a proposal for a new conceptual approach to economic policy is presented. It is driven by convergence and takes into account the general principles of layer-consistent governance, technological neutrality of regulation, and non-discrimination. It is suggested that regulation follows the layer approach, e.g. it be directed to networks, IT services, and digital goods, but not differentiate between different types of networks as the traditional sector approach does. The non-discrimination principle gives reason for network neutrality, which allows differentiation of prices according to quality (bandwidth), but prevents the discrimination of specific content from specific sources by the Internet service provider due to financial reasons. The approach should be put into force internationally within a two-tier system so that general guidelines are introduced supranationally, and specification and implementation are done nationally. The proposal requires the openness of interfaces of networks, digital goods, IT services as far as it is necessary to separate services and content from networks, but the other topics regarding the intellectual property rights of IT services and digital goods remain subject to layer specific solutions. Following this proposal, competition policy will become more important again. The non-discrimination principle supports sovereignty of customers which counterbalances a company's market power, but misuse of market power will be possible in the layer-orientated layer model too. But when the general guidelines, especially network neutrality, are introduced internationally, ex post competition policy is in a good position to prevent the misuse of market power.

Summarizing, digitalization leads to increasing international integration: by fostering the international trade of goods and services, foreign direct investments, the business strategies of the companies involved, and the need to harmonize policy approaches across countries and within sectors. Economic dangers such as monopolization tendencies or criminal threats have to be addressed, but a consistent international policy model will help to fully exploit the benefits of digitalization in open economies.

Chapter 12
Appendices

12.1 Telegraph Lines

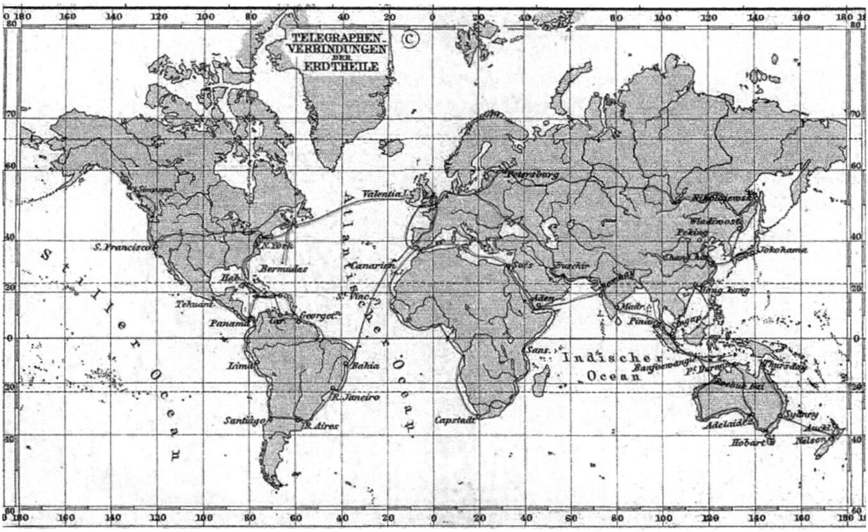

Fig. 12.1 Worldwide telegraph lines in 1891; Original Source: Stieler (1891); Data Source: Wikipedia Commons

M. Vogelsang, *Digitalization in Open Economies*, Contributions to Economics,
DOI 10.1007/978-3-7908-2392-9_12, © Springer-Verlag Berlin Heidelberg 2010

12.2 Sells and Purchases Ordered over the Internet

Table 12.1 Share of companies which use the internet to sell or purchase products; selected countries; Source: OECD Database

OECD key ICT indicators

Internet selling and purchasing (1) by industry (2), 2005

Percentage of businesses with 10 or more employees in each industry group

		Selling	Purchasing
Australia (3)	Construction	20.7	44.0
	Manufacturing	20.6	32.6
	Real estate, renting and business activities	11.1	41.6
	Transport and storage	36.4	49.9
	Communication	35.0	69.9
	Wholesale	25.7	66.4
	Retail	11.7	40.6
	All industries	17.2	45.2
Austria	Construction	4.1	14.5
	Manufacturing	11.6	18.5
	Real estate, renting and business activities	5.5	31.9
	Transport, storage and communication	6.8	14.7
	Wholesale and retail	12.1	25.7
	All industries	10.0	22.2
Belgium	Construction	3.1	10.2
	Manufacturing	20.0	14.8
	Real estate, renting and business activities	11.7	22.1
	Transport, storage and communication	20.8	14.8
	Wholesale and retail	18.5	22.7
	All industries	16.4	17.7
Canada	Construction	0.2	46.3
	Manufacturing	10.2	61.2
	Real estate, renting and business activities	8.6	61.3
	Transport, storage and communication	6.9	49.5
	Wholesale	17.0	64.7
	Retail	14.3	57.8
	All industries	10.7	60.8
Czech Republic	Construction	6.8	16.4
	Manufacturing	14.2	18.4
	Real estate, renting and business activities	12.4	30.5
	Transport, storage and communication	16.3	22.6
	Wholesale and retail	12.9	20.1
	All industries	13.1	21.0
Denmark	All industries	32.0	32.2
Finland	All industries	17.4	19.1

(continued)

Table 12.1 (continued)

		Selling	Purchasing
Germany	Construction	5.0	28.7
	Manufacturing	17.6	37.7
	Real estate, renting and business activities	10.2	50.9
	Transport, storage and communication	18.2	37.2
	Wholesale and retail	20.7	45.5
	All industries	16.0	41.5
Greece	Construction	1.7	14.8
	Manufacturing	4.1	10.7
	Real estate, renting and business activities	5.1	23.5
	Transport, storage and communication	8.3	8.2
	Wholesale and retail	4.9	15.3
	All industries	6.8	13.9
Hungary	Construction	1.1	1.2
	Manufacturing	4.8	4.0
	Real estate, renting and business activities	1.6	4.5
	Transport, storage and communication	2.3	3.0
	Wholesale and retail	4.3	7.8
	All industries	3.9	4.7
Iceland (2003)	Construction	0.0	12.1
	Manufacturing	9.3	8.7
	Real estate, renting and business activities	10.6	25.0
	Transport, storage and communication	17.0	17.0
	Wholesale and retail	18.3	20.2
	All industries	11.8	15.2
Ireland	Construction	6.6	40.5
	Manufacturing	27.8	42.5
	Real estate, renting and business activities	14.8	47.7
	Transport, storage and communication	29.4	37.3
	Wholesale and retail	18.3	41.3
	All industries	21.0	40.6
Italy	Construction	0.7	2.6
	Manufacturing	1.7	1.7
	Real estate, renting and business activities	1.3	12.6
	Transport, storage and communication	2.9	5.1
	Wholesale and retail	2.8	7.1
	All industries	2.7	4.2

(continued)

Table 12.1 (continued)

		Selling	Purchasing
Korea	Construction	9.1	28.4
	Manufacturing	8.0	23.4
	Real estate, renting and business activities	5.3	28.7
	Transport and communication	9.4	29.2
	Wholesale	15.2	30.3
	Retail	12.9	23.1
	All industries	7.7	29.5
Netherlands	Construction	3.4	16.0
	Manufacturing	17.9	16.7
	Real estate, renting and business activities	11.0	26.8
	Transport, storage and communication	21.3	14.6
	Wholesale and retail	15.6	20.9
	All industries	14.2	20.3
Norway	Construction	11.8	31.1
	Manufacturing	26.9	33.2
	Real estate, renting and business activities	22.8	51.3
	Transport, storage and communication	22.1	33.7
	Wholesale and retail	30.4	32.8
	All industries	25.7	36.0
Poland	Construction	0.9	5.0
	Manufacturing	5.2	6.8
	Real estate, renting and business activities	2.3	11.1
	Transport, storage and communication	7.9	9.4
	Wholesale and retail	5.2	13.7
	All industries	4.6	9.4
Portugal	All industries	8.6	12.4
Slovak Republic	Construction	3.0	3.4
	Manufacturing	7.0	4.8
	Real estate, renting and business activities	1.5	16.8
	Transport, storage and communication	4.4	5.6
	Wholesale and retail	10.3	5.8
	All industries	6.8	6.8
Spain	Construction	0.0	1.4
	Manufacturing	3.7	1.9
	Real estate, renting and business activities	1.8	7.2
	Transport, storage and communication	1.2	3.3
	Wholesale and retail	3.4	6.4
	All industries	2.5	3.7

(continued)

Table 12.1 (continued)

		Selling	Purchasing
Sweden	Construction	9.1	28.7
	Manufacturing	24.6	34.7
	Real estate, renting and business activities	20.0	55.7
	Transport, storage and communication	19.5	29.8
	Wholesale and retail	29.4	45.3
	All industries	23.2	41.1
United Kingdom	Construction	8.9	46.0
	Manufacturing	37.3	49.8
	Real estate, renting and business activities	20.0	56.7
	Transport, storage and communication	20.7	50.2
	Wholesale and retail	28.6	43.9
	All industries	25.2	50.8

1. The definition of Internet selling and purchasing varies between countries, with some explicitly including orders placed by conventional email (for instance, Australia and Canada) and others explicitly excluding such orders (e.g. Ireland, the UK and some other European countries). Most countries explicitly use the OECD concept of Internet commerce, that is, goods or services are ordered over the Internet but payment and/or delivery may be off-line.
2. All industries includes: for most European countries: Manufacturing, Construction, Wholesale and retail, (part) Hotels and restaurants, Transport, storage and communication, Real estate, renting and business activities and (part) Other community, social and personal service activities. For Australia, Agriculture, forestry and fishing, Education and Religious organisations are excluded. For Canada, Agriculture, fishing, hunting and trapping, and Construction – specialist contractors are excluded. For Mexico, data refer to enterprises with 50 or more employees and include: Manufacturing, Services and Construction.
3. Data relates to 2004–05 financial year. Internet income results from orders received via the Internet or Web for goods or services, where an order is a commitment to purchase.

Source: OECD, ICT database and Eurostat, Community Survey on ICT usage in enterprises, September 2006.

12.3 Exports and Communication: An Empirical Analysis

The model of Chap. 4 states a positive relationship between communication and trade. The goal of this appendix is to analyze, whether this thesis can be confirmed or not by the data.

Not all regressions are shown here. A summary of the results is presented in Table 12.3. The following data have been used:

- experc: Exports of goods and services (% of GDP); 2005; source: World Development Indicators (WDI)
- gdpabs: GDP Current USD 2005; source: WDI
- extimesgdp: Exports in current USD
- intu: Internet Users per 1000, 2005; source: WDI
- pop: Population Total, 2005; source: WDI
- bandwto: International Internet Bandwidth (Mbps), 2005; source: International Telecommunication Union (ITU)

- bandwinh: International Internet Bandwidth per inhabitant (bit/s), 2005; source: ITU
- telout: International outgoing fixed telephone traffic (minutes), 2005; source: ITU
- internu2: Internet users, 2005; source: ITU
- arbeitsk: Hourly compensation costs of production workers in manufacturing in US Dollars, 2005; source: International Labour Organization (ILO)
- labunit: Labour compensation per unit of output (1997 USD at constant PPP), 2005; source: ILO

After having cleared the data-set and amendments of some missing data for single countries (additional sources: OECD, U.S. Department of Commerce, Bureau of Economic Analysis) the data set lists 131 countries. Unfortunately the availability of labour compensation and unit costs is poor: There are only 24 countries with report data for the variables labunit and bandwto. This constrains the comparability of the regressions.

12.3.1 Descriptive Analysis

The scatter diagram in Fig. 12.2 shows the relationship between outgoing telephone minutes and total exports graphically. The correlation matrix in Table 12.2 points to the high correlation between outgoing telephone minutes and bandwidth:

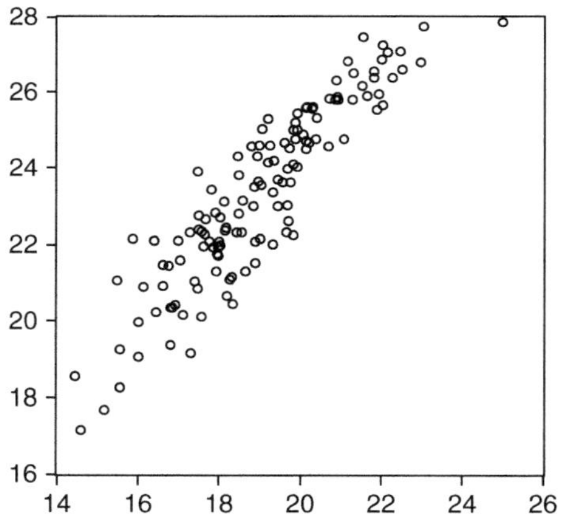

Fig. 12.2 Scatter diagram: outgoing telephone minutes (log) and total exports (log)

Table 12.2 Correlation matrix

	LOG(TELOUT)	LOG(BANDWTO)	LOG(POP)
LOG(TELOUT)	1.000	0.843	0.538
LOG(BANDWTO)	0.843	1.000	0.449
LOG(POP)	0.538	0.449	1.000

12.3.2 Econometric Analysis I: Estimation for Total Exports

First the total exports of a country are estimated. Because the variable extimesgdp varies extremely with country size, the size-effect should be captured by pop. This leads to estimation 1:

$$log(exptimesgdp) = c + \beta_1 log(pop) + \beta_2 log(telout) + \beta_3 log(bandwto)$$

Null hypothesis: $\beta_2 = 0; \beta_3 = 0$, Expectation (alternative hypothesis): $\beta_2 > 0; \beta_3 > 0$

Result of Fig. 12.3: When the internet bandwidth rise 1%, then this contributes in 0.2 percent increase in the export volume ceteric paribus.

But it is to fear, that heteroskedasticity may exist: Therefore the White test is done:

The null hypothesis is homoskedasticity: The test statistic amounts $NR^2 = 122 * 0.062 = 7.56$. The critical value of the chi-square distribution is > 50, so that the null-hypothesis cannot be rejected. The White test shows no indication for heteroskedasticity in estimation 1.

Another test of heteroskedasticity is the park test:

The logarithm of the population has a small but significant influence on the square of the residuals. This might be interpreted as a sign for heteroskedasticity. Summarizing the results of Park and White tests the null-hypothesis of heteroskedasticity cannot be rejected.

Multicollinearity might be also a severe impairment of estimation 1. As a rule of the thumb the Variance Inflation factors (VIF)

$$VIF = \frac{1}{1 - R^2}$$

should be smaller than 5.

Dependent Variable: LOG(EXPTIMESGDP)
Method: Least Squares
Date: 02/27/08 Time: 16:40
Sample: 1 131
Included observations: 128
Excluded observations: 3

Variable	Coefficient	Std. Error	t-Statistic	Prob.
C	5.375765	1.146282	4.689738	0.0000
LOG(POP)	0.269968	0.046521	5.803091	0.0000
LOG(TELOUT)	0.640727	0.073701	8.693646	0.0000
LOG(BANDWTO)	0.217151	0.038737	5.605837	0.0000

R-squared	0.883431	Mean dependent var	23.24503
Adjusted R-squared	0.880611	S.D. dependent var	2.265884
S.E. of regression	0.782925	Akaike info criterion	2.379192
Sum squared resid	76.00850	Schwarz criterion	2.468318
Log likelihood	-148.2683	F-statistic	313.2493
Durbin-Watson stat	2.137728	Prob(F-statistic)	0.000000

Fig. 12.3 Estimation 1

Dependent Variable: SCHAETZ1RESID^2				
Method: Least Squares				
Date: 02/27/08 Time: 17:07				
Sample: 1 131				
Included observations: 128				
Excluded observations: 3				
Variable	Coefficient	Std. Error	t-Statistic	Prob.
C	−7.392768	16.48433	−0.448472	0.6546
LOG(POP)	0.063391	1.066623	0.059431	0.9527
LOG(TELOUT)	0.895276	2.054277	0.435811	0.6638
LOG(BANDWTO)	−0.037997	0.967905	−0.039257	0.9688
LOG(POP)^2	−0.004060	0.026645	−0.152371	0.8792
LOG(TELOUT)^2	−0.034765	0.075869	−0.458221	0.6476
LOG(BANDWTO)^2	0.011131	0.020175	0.551727	0.5822
LOG(POP)*LOG(TELOUT)	0.014028	0.065373	0.214587	0.8305
LOG(POP)*LOG(BANDWTO)	−0.032854	0.036395	−0.902723	0.3685
LOG(TELOUT)*LOG(BANDWTO)	0.019815	0.069345	0.285754	0.7756
R-squared	0.128224	Mean dependent var		0.593816
Adjusted R-squared	0.061733	S.D. dependent var		0.905480
S.E. of regression	0.877086	Akaike info criterion		2.650481
Sum squared resid	90.77502	Schwarz criterion		2.873296
Log likelihood	−159.6308	F-statistic		1.928434
Durbin-Watson stat	2.025380	Prob(F-statistic)		0.054254

Fig. 12.4 White test

Dependent Variable: LOG(SCHAETZ1RESID^2)				
Method: Least Squares				
Date: 02/27/08 Time: 17:04				
Sample: 1 131				
Included observations: 128				
Excluded observations: 3				
Variable	Coefficient	Std. Error	t-Statistic	Prob.
C	1.949874	2.069561	0.942168	0.3479
LOG(POP)	−0.256032	0.128714	−1.989161	0.0488
R-squared	0.030447	Mean dependent var		−2.141961
Adjusted R-squared	0.022752	S.D. dependent var		2.598713
S.E. of regression	2.568980	Akaike info criterion		4.740396
Sum squared resid	831.5568	Schwarz criterion		4.784959
Log likelihood	−301.3854	F-statistic		3.956761
Durbin-Watson stat	2.212010	Prob(F-statistic)		0.048850

Fig. 12.5 Park test

It results a $VIF(\beta_2) = 3.85$ (not shown in the regression result table), and $VIF(\beta_3) = 3,4$ (figure 12.6). It seems that no severe multicollinearity is at work.

12.3.3 Econometric Analysis II: Estimation for Export Share

The 2nd approach estimates the export shares:

$$experc = c + \beta_2(telout/pop) + \beta_3(bandwto/pop)$$

Dependent Variable: LOG(BANDWTO)
Method: Least Squares
Date: 02/27/08 Time: 17:34
Sample: 1 131
Included observations: 128
Excluded observations: 3

Variable	Coefficient	Std. Error	t-Statistic	Prob.
C	−21.93987	1.776083	−12.35296	0.0000
LOG(POP)	−0.010331	0.107414	−0.096184	0.9235
LOG(TELOUT)	1.518411	0.102541	14.80783	0.0000

R-squared	0.710139	Mean dependent var	6.613784
Adjusted R-squared	0.705501	S.D. dependent var	3.331204
S.E. of regression	1.807769	Akaike info criterion	4.045223
Sum squared resid	408.5038	Schwarz criterion	4.112067
Log likelihood	−255.8943	F-statistic	153.1205
Durbin-Watson stat	1.826339	Prob(F-statistic)	0.000000

Fig. 12.6 Regression for variance inflation factor

Dependent Variable: EXPERC
Method: Least Squares
Date: 06/01/08 Time: 17:58
Sample(adjusted): 1 131
Included observations: 29
Excluded observations: 102

Variable	Coefficient	Std. Error	t-Statistic	Prob.
C	43.63843	9.570441	4.559710	0.0001
TELOUT/POP	0.179061	0.021873	8.186450	0.0000
BANDWTO/POP	1176.021	839.9331	1.400136	0.1738
ARBEITSK	−1.386155	0.525590	−2.637330	0.0142
R-squared	0.796742	Mean dependent var		
Adjusted R-squared	0.772351	S.D. dependent var		
S.E. of regression	25.34737	Akaike info criterion		
Sum squared resid	16062.22	Schwarz criterion		
Log likelihood	−132.7447	F-statistic		
Durbin-Watson stat	2.547267	Prob(F-statistic)		

Fig. 12.7 Estimation 2b

Null hypothesis: $\beta_2 \leq= 0; \beta_3 \leq= 0$

This estimation (2a - not shown here) shows a positive impact of telout/pop and bandwto/pop on the export share, but only the outgoing telephone minutes per head are significant. Of course the problem of omitted variables is existent as labour cost variables are expected to have an impact on export shares too.

Estimation 2b (Fig. 12.7) accounts for labour costs. The sign are as expected, the influence of labour costs is significant. The adjustment of labour costs for productivity, so that variable labunit is used instead of arbeitsk, does not change the results.

Table 12.3 Summary of Estimations

	Outgoing telephone minutes	Internet bandwidth	Internet user
1. Exports in USD, absolute	Absolute positive, significant	Absolute positive, significant	Absolute positive, not sign
2. Export share (in % of GDP)	Per head positive, signficant	Per head positive, not sign	Per head chang. sign

12.3.4 Interpretation

Table 12.3 shows that the influence of telephone minutes for exports is robust. This is a welcome result, as the fundamental interpretation between Sect. 12.3.2 and 12.3.3 changes: With estimation 1 it can be argued, that a large country have more outgoing telephone calls and more total exports. With estimation 2 a counterbalancing effect could be true: Small countries have a larger degree of openness and therefore have more outgoing telephone minutes per head and higher export shares. Summarizing:

- The fundamental conclusion of Chap. 4 cannot be rebutted by the data.

That the impact of internet bandwidth on trade is not so clear does not contradict this result: To organize trade internationally, e.g. via emails, only a small bandwidth is necessary. Also further estimations (here not shown) show that a positive relationship between internet variables (bandwidth, users) and per-head income cannot be confirmed with this data-set: Korea, Slovenia, Estland have relatively small income figures, but high internet penetration rates, whereas inhabitants of oil exporting countries such as Kuwait have a relatively high income, but use the internet less.

12.4 Variables of the IT Services Model

Most important variables used in Chap. 5:

A_Y	Parameter; real expenditures in sector Y
α	Coefficient; $1/(1-\alpha)$ elasticity of substitution (constant)
β	Cobb-Douglas production coefficient; $0 < \beta_Z < \beta_Y < 1$; Industry Y is more capital intensive than industry Z
$\widehat{\beta}$	Knife-edge case; outsourcing as well as integration in a sector
D	Costs for IT Services (fix costs) per variant of the final good
E	Total expenditures
ε_k	Parameter
k	Index for sectors Y and Z
δ	In the case of integration: share of residual rights which are hold by final product producer; $0 < \delta < 1$

$K_{x,k}$(i)	Capital invested for intermediates x(i) of sector k
K^*	Factor endowment with capital
K_Y^d	Total demand for capital of sector Y
L	Labour; use of indices analogue to K
l	Index, $l \in \{I, S\}$
I	Subindex, indicates the case of integrated companies
n_y	Number of variants in sector Y
$n_{Y,I}$	Pairs of integrated firms in sector Y
μ	Share of total consumption for final products of sector Y; constant
Ω^A	Capital-labour endowment of country A
$p_{Y,I}(i)$	Price of variant i of final product of integrated producer
$p_{Y,S}(i)$	Price of variant i of final product of non-integrated producer
ϕ	Nash-Bargaining share of final product producer; assumption: $1/2 < \phi < 1$
$\overline{\phi}$	Parameter; $\overline{\phi} = \delta^\alpha + \phi(1 - \delta^\alpha) > \phi$
S	Index: Separated Companies (Outsourcing-Solution)
R_Y	Potential revenues from sale of a final product in sector Y
r	Interest rate
σ_L	Average labour share in economy
$T_k(i)$	Lump-sum transfer from intermediate producer to final product producer
$x_Y(i)$	Intermediates for variant i of final product of sector Y
w	Wage
y(i)	Variants of final product of sector Y
z(i)	Variants of final product of sector Z
ζ	Share of endowment of world economy used for IT Services

12.5 Calculation of the Demand Function for y(i)

The preferences of a representative consumer are:

$$U = \left(\int_0^{n_y} y(i)^\alpha di \right)^{\mu/\alpha} \left(\int_0^{n_z} z(i)^\alpha di \right)^{\frac{1-\mu}{\alpha}} \tag{12.5.1}$$

with elasticity of substitution between any two varieties in a given sector

$$\frac{1}{1-\alpha} > 1$$

and

$$0 < \mu < 1$$

With

$$E = \int_0^{n_y} (p_y(i)y(i))di + \int_0^{n_z} (p_z(i)z(i))di \tag{12.5.2}$$

with

$$E_Y = \mu E = \int_0^{n_y} (p_y(i)y(i))di \tag{12.5.3}$$

for spending for products in sector Y and for sector Z respectively. By assumption, in equilibrium spending for sector Y and use of products of sector Y both have a price-adjusted share of μ of total spending or use respectively. Mathematically this is a result of the Cobb-Douglas type demand function[1]

$$L = \left(\int_0^{n_y} y(i)^\alpha di\right)^{\mu/\alpha} \left(\int_0^{n_z} z(i)^\alpha di\right)^{\frac{1-\mu}{\alpha}}$$
$$-\lambda\left(E - \int_0^{n_y} (p_y(i)y(i))di + \int_0^{n_z} (p_z(i)z(i))di\right) \tag{12.5.4}$$

Calculating the derivates and setting them zero leads to[2]

$$\frac{\partial L}{\partial y(i)} = \alpha\frac{\mu}{\alpha}y(i)^{\alpha-1}\left(\int_0^{n_y} y(i)^\alpha di\right)^{\mu/\alpha-1}\left(\int_0^{n_z} z(i)^\alpha di\right)^{\frac{1-\mu}{\alpha}} n_y - \lambda p_y(i)n_y = 0 \tag{12.5.5}$$

$$\frac{\partial L}{\partial z(i)} = \alpha\frac{1-\mu}{\alpha}z(i)^{\alpha-1}\left(\int_0^{n_y} y(i)^\alpha di\right)^{\mu/\alpha}\left(\int_0^{n_z} z(i)^\alpha di\right)^{\frac{1-\mu}{\alpha}-1} n_z - \lambda p_z(i)n_z = 0 \tag{12.5.6}$$

and

$$\frac{\partial L}{\partial \lambda} = \int (p_y(i)y(i))di + \int (p_z(i)z(i))di - E =!0 \tag{12.5.7}$$

The optimization procedure is showed for Y in the following. Rearranging (12.5.5) gives:

$$y(i) = \frac{1}{\mu}^{\frac{1}{1-\alpha}}\left(\int_0^{n_y} y(i)^\alpha di\right)^{\frac{\alpha-\mu}{\alpha}\frac{1}{\alpha-1}}\left(\int_0^{n_z} z(i)^\alpha di\right)^{\frac{1-\mu}{\alpha}\frac{1}{\alpha-1}} \lambda^{\frac{1}{\alpha-1}} p_y(i)^{\frac{1}{\alpha-1}} \tag{12.5.8}$$

Inserting this equation into (12.5.7) for all y(i) leads to:

$$E = \int_0^{n_y} p_y(i)\left(\frac{1}{\mu}^{\frac{1}{1-\alpha}}\left(\int_0^{n_y} y(i)^\alpha di\right)^{\frac{\alpha-\mu}{\alpha}\frac{1}{\alpha-1}}\left(\int_0^{n_z} z(i)^\alpha di\right)^{\frac{1-\mu}{\alpha}\frac{1}{\alpha-1}}\right.$$
$$\left.\lambda^{\frac{1}{\alpha-1}} p_y(i)^{\frac{1}{\alpha-1}}\right) di + \int_0^{n_z} (p_z(i)z(i))$$

---

[1] Proof: Optimization of $U = Y/muZ^{1-\mu}$ with budget restriction $E = p_Y Y p_Z Z$ leads to $\frac{p_Z \mu Z}{p_Y(1-\mu)Y} = 1$ with can be rearranged with the budget restriction to $Y = \frac{\mu E}{p_Y}$ and $Z = \frac{(1-\mu)E}{p_Z}$. Thus the demand for Y only depends on share μ of total spending E and the exogenous price for this product. Demand of Z and price of Z do not influence the demand for Y, what is a standard result of a Cobb-Douglas type demand function with constant elasticities.

[2] Please note, that y(i) does not describe a functional dependence, but i acts as an index from mathematical point of view.

Multiplying the $p_y(i)$ and rearranging this equation results in:

$$\frac{E - \int_0^{n_z}(p_z(i)z(i))di}{\int_0^{n_y} p_y(i)^{\frac{\alpha}{\alpha-1}}di} = \frac{1}{\mu}^{\frac{1}{1-\alpha}} \lambda^{\frac{1}{\alpha-1}} \left(\int_0^{n_y} y(i)^{\alpha} di\right)^{\frac{\alpha-\mu}{\alpha}\frac{1}{\alpha-1}} \left(\int_0^{n_z} z(i)^{\alpha} di\right)^{\frac{1-\mu}{\alpha}\frac{1}{\alpha-1}}$$

Introducing the right side into (12.5.8) amounts to:

$$y(i) = \frac{E - \int_0^{n_z}(p_z(i)z(i))di}{\int_0^{n_y} p_y(i)^{\frac{\alpha}{\alpha-1}}di} p_y(i)^{\frac{1}{\alpha-1}} \tag{12.5.9}$$

Using 12.5.3 this leads to:

$$y(i) = \frac{\mu E}{\int_0^{n_y} p_y(i)^{\frac{\alpha}{\alpha-1}}di} p_y(i)^{\frac{1}{\alpha-1}} \tag{12.5.10}$$

12.6 Incomplete Contracts

Optimal investment volumes are calculated from (5.15) and (5.16).

According to (5.14) $R_Y(i)$ is replaced and indices are omitted here to simplify the notification.

The task is to maximise

$$\psi A^{1-\alpha}\left(\frac{K}{\beta}\right)^{\alpha\beta}\left(\frac{L}{1-\beta}\right)^{\alpha(1-\beta)} - rK - rK_0 \Rightarrow MAX! \tag{12.6.1}$$

by choosing K and

$$(1-\psi)A^{1-\alpha}\left(\frac{K}{\beta}\right)^{\alpha\beta}\left(\frac{L}{1-\beta}\right)^{\alpha(1-\beta)} - wL - wL_0 \Rightarrow MAX! \tag{12.6.2}$$

by choosing L simultaneously. By assumption interest rate r, wage w and investment for IT services K_0 and L_0 are sufficiently small, so that profit is positive for the investment volumes calculated.

Setting the first derivate of (12.6.1) and (12.6.2) to zero lead to

$$\frac{\partial}{\partial K} = \psi A^{1-\alpha}\alpha\beta\frac{1}{\beta}\left(\frac{K}{\beta}\right)^{\alpha\beta-1}\left(\frac{L}{1-\beta}\right)^{\alpha(1-\beta)} - r = 0 \tag{12.6.3}$$

and

$$\frac{\partial}{\partial L} = (1-\psi)A^{1-\alpha}\left(\frac{K}{\beta}\right)^{\alpha\beta}\frac{\alpha(1-\beta)}{1-\beta}\left(\frac{L}{1-\beta}\right)^{\alpha(1-\beta)-1} - w = 0 \qquad (12.6.4)$$

In order to be maximised, the second derivates should be < 0, which is fulfilled for all K, L by:

$$\frac{\partial^2}{(\partial K)^2} = \psi A^{1-\alpha}\frac{\alpha(\alpha\beta-1)}{\beta}\left(\frac{K}{\beta}\right)^{\alpha\beta-2}\left(\frac{L}{1-\beta}\right)^{\alpha(1-\beta)} < 0 \qquad (12.6.5)$$

with

$$\alpha\beta < 1$$

and

$$\frac{\partial^2}{(\partial L)^2} = (1-\psi)A^{1-\alpha}\left(\frac{K}{\beta}\right)^{\alpha\beta}\frac{\alpha(\alpha(1-\beta)-1)}{1-\beta}\left(\frac{L}{1-\beta}\right)^{\alpha(1-\beta)-2} - w < 0 \qquad (12.6.6)$$

with

$$\alpha < \alpha\beta + 1$$

Equations (12.6.3) and (12.6.4) are two reactions functions, which show the optimal investment of K (L) in dependence on the investment volume of L (K) of the integrated supplier (final goods producer respectively). To find an equilibrium the intersection of the two functions is calculated.

Equation (12.6.3) can also be expressed as:

$$\left(\frac{K}{\beta}\right)^{-\alpha\beta} = \frac{\psi A^{1-\alpha}\frac{\alpha}{r}\left(\frac{L}{1-\beta}\right)^{\alpha(1-\beta)}}{\frac{K}{\beta}} \qquad (12.6.7)$$

or

$$\frac{K}{\beta} = \left(\frac{\psi\alpha}{r}\right)^{\frac{1}{1-\alpha\beta}} A^{\frac{1-\alpha}{1-\alpha\beta}}\left(\frac{L}{1-\beta}\right)^{\frac{\alpha(1-\beta)}{1-\alpha\beta}} \qquad (12.6.8)$$

Equation (12.6.4) can be transformed to:

$$\left(\frac{K}{\beta}\right)^{-\alpha\beta} = (1-\psi)A^{1-\alpha}\frac{\alpha}{w}\left(\frac{L}{1-\beta}\right)^{\alpha(1-\beta)-1} \qquad (12.6.9)$$

or

$$\frac{L}{1-\beta} = \frac{(1-\psi)\alpha}{w}^{\frac{1}{1-\alpha(1-\beta)}} A^{\frac{1-\alpha}{1-\alpha(1-\beta)}}\left(\frac{K}{\beta}\right)^{\frac{\alpha\beta}{1-\alpha(1-\beta)}} \qquad (12.6.10)$$

Equalizing the right sides of (12.6.9) and (12.6.7) and simplifying leads to:

$$\frac{\psi}{1-\psi}\frac{w}{r}\frac{L}{1-\beta} = \frac{K}{\beta} \tag{12.6.11}$$

Introducing (12.6.8) in (12.6.11) then result in:

$$\frac{\psi}{1-\psi}\frac{w}{r}\frac{L}{1-\beta} = \psi^{\frac{1}{1-\alpha\beta}} A^{\frac{1-\alpha}{1-\alpha\beta}} \left(\frac{\alpha}{r}\right)^{\frac{1}{1-\alpha\beta}} \left(\frac{L}{1-\beta}\right)^{\frac{\alpha(1-\beta)}{1-\alpha\beta}}$$

Rearranging leads to:

$$\psi^{1-\frac{1}{1-\alpha\beta}} \frac{1}{1-\psi}\frac{w}{r} A^{\frac{\alpha-1}{1-\alpha\beta}} \left(\frac{\alpha}{r}\right)^{\frac{-1}{1-\alpha\beta}} = \left(\frac{L}{1-\beta}\right)^{\frac{\alpha(1-\beta)}{1-\alpha\beta}-1} = \left(\frac{L}{1-\beta}\right)^{\frac{\alpha-1}{1-\alpha\beta}}$$

Simplifying result in:

$$L = A(1-\beta)\psi^{\frac{-\alpha\beta}{\alpha-1}} (1-\psi)^{\frac{\alpha\beta-1}{\alpha-1}} w^{\frac{1-\alpha\beta}{\alpha-1}} r^{\frac{\alpha\beta}{\alpha-1}} \alpha^{\frac{-1}{\alpha-1}} \tag{12.6.12}$$

$$L = A(1-\beta)\left(\alpha\psi^{\alpha\beta}(1-\psi)^{1-\alpha\beta} w^{\alpha\beta-1} r^{-\alpha\beta}\right)^{\frac{1}{1-\alpha}} \tag{12.6.13}$$

Solving for K (12.6.10) is put into (12.6.11) leading to:

$$\frac{\psi}{1-\psi}\frac{w}{r}\left(\frac{(1-\psi)\alpha}{w}\right)^{\frac{1}{1-\alpha(1-\beta)}} A^{\frac{1-\alpha}{1-\alpha(1-\beta)}} \left(\frac{K}{\beta}\right)^{\frac{\alpha\beta}{1-\alpha(1-\beta)}} = \frac{K}{\beta}$$

Finally rearranging and simplifying leads to:

$$K = \beta A \psi^{\frac{1-\alpha(1-\beta)}{1-\alpha}} (1-\psi)^{\frac{\alpha(1-\beta)}{1-\alpha}} r^{\frac{\alpha(1-\beta)-1}{1-\alpha}} w^{\frac{-\alpha(1-\beta)}{1-\alpha}} \alpha^{\frac{1}{1-\alpha}} \tag{12.6.14}$$

$$K = \beta A\left(\psi^{1-\alpha(1-\beta)}(1-\psi)^{\alpha(1-\beta)} r^{\alpha(1-\beta)-1} w^{-\alpha(1-\beta)}\alpha\right)^{\frac{1}{1-\alpha}} \tag{12.6.15}$$

12.7 Complete Contracts

The task is to maximise

$$\phi\left[A^{1-\alpha}\left(\frac{K}{\beta}\right)^{\alpha\beta}\left(\frac{L}{1-\beta}\right)^{\alpha(1-\beta)} - rK - wL - rK_0 - wL_0\right] \Rightarrow MAX!$$

by choosing K and

$$(1-\phi)\left[A^{1-\alpha}\left(\frac{K}{\beta}\right)^{\alpha\beta}\left(\frac{L}{1-\beta}\right)^{\alpha(1-\beta)}-rK-wL-rK_0-wL_0\right]\Rightarrow MAX!$$

by choosing L simultaneously. Because a constant $\phi > 0$ does not influence the optimization result, the task reduces to

$$A^{1-\alpha}\left(\frac{K}{\beta}\right)^{\alpha\beta}\left(\frac{L}{1-\beta}\right)^{\alpha(1-\beta)}-rK-wL-rK_0-wL_0\Rightarrow MAX! \quad (12.7.1)$$

Following the calculations of Sect. 12.6 this leads to

$$\frac{w}{r}\frac{L}{1-\beta}=\frac{K}{\beta} \quad (12.7.2)$$

The factor demands in the case of complete contracts are:

$$L = A(1-\beta)\left(\alpha w^{\alpha\beta-1}r^{-\alpha\beta}\right)^{\frac{1}{1-\alpha}} \quad (12.7.3)$$

and

$$K = \beta A\left(r^{\alpha(1-\beta)-1}w^{-\alpha(1-\beta)}\alpha\right)^{\frac{1}{1-\alpha}} \quad (12.7.4)$$

12.7.1 Calculation of First Derivate of θ with Regard to β

The task is to calculate the first derivative of (5.33) and to prove that it is > 0.

$$\theta=\left(1+\frac{\alpha(1-2\beta)\delta^{\alpha}(1-\phi)}{1-\alpha+\alpha\beta-(2\alpha\beta-\alpha)\phi}\right)\left(1+\frac{\delta^{\alpha}}{\phi(1-\delta^{\alpha})}\right)^{\frac{\alpha\beta}{1-\alpha}}(1-\delta^{\alpha})^{\frac{\alpha}{1-\alpha}}>0 \quad (12.7.5)$$

with regard to β.

To simplify notification I define the functions

$$f(\beta)=\left(1+\frac{\alpha(1-2\beta)\delta^{\alpha}(1-\phi)}{1-\alpha+\alpha\beta-(2\alpha\beta-\alpha)\phi}\right) \quad (12.7.6)$$

$$g(\beta)=\left(1+\frac{\delta^{\alpha}}{\phi(1-\delta^{\alpha})}\right)^{\frac{\alpha\beta}{1-\alpha}}(1-\delta^{\alpha})^{\frac{\alpha}{1-\alpha}} \quad (12.7.7)$$

$$h(\beta)=\alpha(1-2\beta)\delta^{\alpha}(1-\phi) \quad (12.7.8)$$

$$i(\beta)=1-\alpha+\alpha\beta-(2\alpha\beta-\alpha)\phi \quad (12.7.9)$$

$$B = \frac{\delta^\alpha}{\phi(1-\delta^\alpha)} \tag{12.7.10}$$

Applying the product rule, the first derivative of (12.7.5) is given by

$$\frac{\partial\theta}{\partial\beta} = \theta' = f'g + g'f = (h'[i]^{-1} - hi'[i]^{-2})g + g'f \tag{12.7.11}$$

with

$$h' = -2\alpha(1-\phi)\delta^\alpha \tag{12.7.12}$$

$$i' = \alpha - 2\alpha\beta \tag{12.7.13}$$

$$g' = (1+B)^{\frac{\alpha\beta}{1-\alpha}} \frac{\alpha}{1-\alpha} ln(1+B)(1-\delta^\alpha)^{\frac{\alpha}{1-\alpha}} \tag{12.7.14}$$

The first summand of the derivative $\frac{\partial\theta}{\partial\beta}$ is

$$\begin{aligned} f'g &= \frac{h'i - i'h}{i^2} \\ &= \frac{a(1-\phi)\delta^\alpha[-2-2\alpha-2\alpha\beta-2\alpha\phi+4\alpha\phi\beta] - [\alpha-2\alpha\phi-2\alpha\beta+4\alpha\beta\phi]}{i^2} \end{aligned} \tag{12.7.15}$$

This reduces to

$$f'g = \frac{(1-\phi)\delta^\alpha(-2\alpha+\alpha^2)}{i^2} \tag{12.7.16}$$

so that

$$\frac{\partial\theta}{\partial\beta} = \left[\frac{(1-\phi)\delta^\alpha(-2\alpha-\alpha^2)}{i^2} + \left[1+\frac{h}{i}\right]\frac{\alpha}{1-\alpha} ln 1+B\right](1+B)^{\frac{\alpha\beta}{1-\alpha}}(1-\delta^\alpha)^{\frac{\alpha}{1-\alpha}} > 0 \tag{12.7.17}$$

Dividing by the constant factors $(1+B)^{\frac{\alpha\beta}{1-\alpha}}(1-\delta^\alpha)^{\frac{\alpha}{1-\alpha}} > 0$ and subcontracting the first summand gives:

$$\left[1+\frac{h}{i}\right]\frac{\alpha}{1-\alpha} ln(1+B) > \frac{(1-\phi)\delta^\alpha(2\alpha-\alpha^2)}{i^2} \tag{12.7.18}$$

or

$$ln(1+B) + \frac{hiln(1+B)}{i^2} > \frac{(1-\phi)\delta^\alpha(2-\alpha)(1-\alpha)}{i^2} \tag{12.7.19}$$

This relationship is surely satisfied when

$$hiln(1+B) > (1-\phi)\delta^\alpha(2-\alpha)(1-\alpha) \tag{12.7.20}$$

Defining

$$\Omega(\beta) = h(\beta)i(\beta) \tag{12.7.21}$$

leads to

$$\Omega(\beta)ln\left(1+\frac{\delta^{\alpha}}{\phi(1-\delta^{\alpha})}\right) > (1-\phi)\delta^{\alpha}(2-\alpha)(1-\alpha) \tag{12.7.22}$$

with

$$\Omega(\beta) = (1-\alpha(1-\overline{\phi})+\alpha\beta(1-2\overline{\phi}))(1-\alpha(1-\phi)+\alpha\beta(1-2\phi)) \tag{12.7.23}$$

using the calculations for (5.33), esp. the definition $\overline{\phi} = \delta^{\alpha} + \phi(1-\delta^{\alpha})$.

12.7.2 Number of Variants Produced in a Country

$$\begin{aligned} K^A &= \frac{n_Y^A}{n_Y^*}\left[\mu(rK^*+wL^*)\alpha\beta_Y\frac{\overline{\phi}}{r} + \mu K^*(1-\alpha+\alpha\beta_Y - 2\alpha\beta_Y\overline{\phi}+\alpha\overline{\phi})\right] \\ &+\frac{n_Z^A}{n_Z^*}\Bigg[(1-\mu)(rK^*+wL^*)\alpha\beta_Z\frac{\phi}{r} + (1-\mu) \\ &\quad K^*(1-\alpha+\alpha\beta_Z-2\alpha\beta_Z\phi+\alpha\phi)\Bigg] \end{aligned} \tag{12.7.24}$$

From (5.40) and (5.50)

$$\begin{aligned} L^A &= \frac{n_Y^A}{n_Y^*}\Bigg[\mu(rK^*+wL^*)\frac{\alpha(1-\beta_Y)(1-\overline{\phi})}{w} \\ &\quad + \mu L^*(1-\alpha+\alpha\beta_Y-2\alpha\beta_Y\overline{\phi}+\alpha\overline{\phi})\Bigg] \\ &+\frac{n_Z^A}{n_Z^*}\Bigg[(1-\mu)(rK^*+wL^*)\frac{\alpha(1-\beta_Z)(1-\phi)}{w} \\ &\quad +(1-\mu)L^*(1-\alpha+\alpha\beta_Z-2\alpha\beta_Z\phi+\alpha\phi)\Bigg] \end{aligned}$$

Defining

$$\varepsilon_Z = 1-\alpha+\alpha\beta_Z-2\alpha\beta_Z\phi+\alpha\phi \tag{12.7.25}$$

$$\varepsilon_Y = 1-\alpha+\alpha\beta_Y-2\alpha\beta_Y\overline{\phi}+\alpha\overline{\phi} \tag{12.7.26}$$

and

$$\xi = (1+\frac{wL^*}{rK^*}) \tag{12.7.27}$$

$$\frac{\xi}{\xi - 1} = (1 + \frac{rK^*}{wL^*}) \tag{12.7.28}$$

and dividing through K^* and L^* respectively leads to:

$$\begin{aligned}\frac{K^A}{K^*} &= \frac{n_Y^A}{n_Y^*}\left[\mu\xi\alpha\beta_Y\overline{\phi} + \mu\epsilon_Y\right] \\ &+ \frac{n_Z^A}{n_Z^*}\left[(1-\mu)\xi\alpha\beta_Z\phi + (1-\mu)\varepsilon_Z\right] \end{aligned} \tag{12.7.29}$$

$$\begin{aligned}\frac{L^A}{L^*} &= \frac{n_Y^A}{n_Y^*}\left[\mu\frac{\xi}{\xi-1}\alpha(1-\beta_Y)(1-\overline{\phi}) + \mu\varepsilon_Y\right] \\ &+ \frac{n_Z^A}{n_Z^*}\left[(1-\mu)\frac{\xi}{\xi-1}\alpha(1-\beta_Z)(1-\phi) + (1-\mu)\varepsilon_Z\right] \end{aligned} \tag{12.7.30}$$

Solving (12.7.29) for n_Y^A/n_Y^*:

$$\frac{n_Y^A}{n_Y^*} = \frac{\frac{K^A}{K^*} - \frac{n_Z^A}{n_Z^*}\left[(1-\mu)\xi\alpha\beta_Z\phi + (1-\mu)\varepsilon_Z\right]}{\left[\mu\xi\alpha\beta_Y\overline{\phi} + \mu\varepsilon_Y\right]} \tag{12.7.31}$$

Plugging into L^A/L^*:

$$\begin{aligned}\frac{L^A}{L^*} &= \frac{K^A}{K^*}\frac{\left[\mu\frac{\xi}{\xi-1}\alpha(1-\beta_Y)(1-\overline{\phi}) + \mu\varepsilon_Y\right]}{\left[\mu\xi\alpha\beta_Y\overline{\phi} + \mu\varepsilon_Y\right]} \\ &+ \frac{n_Z^A}{n_Z^*}\left(\frac{-\left[(1-\mu)\xi\alpha\beta_Z\phi + (1-\mu)\varepsilon_Z\right]\left[\mu\frac{\xi}{\xi-1}\alpha(1-\beta_Y)(1-\overline{\phi}) + \mu\varepsilon_Y\right]}{\left[\mu\xi\alpha\beta_Y\overline{\phi}+\mu\varepsilon_Y\right]}\right) \\ &+ \frac{n_Z^A}{n_Z^*}\left((1-\mu)\frac{\xi}{\xi-1}\alpha(1-\beta_Z)(1-\phi) + (1-\mu)\varepsilon_Z\right) \end{aligned} \tag{12.7.32}$$

and multiplying with $\left[\mu\xi\alpha\beta_Y\overline{\phi} + \mu\varepsilon_Y\right]$ leads to

$$\begin{aligned}&\frac{L^A}{L^*}\left[\mu\xi\alpha\beta_Y\overline{\phi} + \mu\varepsilon_Y\right] - \frac{K^A}{K^*}\left[\mu\frac{\xi}{\xi-1}\alpha(1-\beta_Y)(1-\overline{\phi}) + \mu\varepsilon_Y\right] \\ &= -\frac{n_Z^A}{n_Z^*}\left[(1-\mu)\xi\alpha\beta_Z\phi + (1-\mu)\varepsilon_Z\right]\left[\mu\frac{\xi}{\xi-1}\alpha(1-\beta_Y)(1-\overline{\phi}) + \mu\varepsilon_Y\right] \\ &+ \frac{n_Z^A}{n_Z^*}\left((1-\mu)\frac{\xi}{\xi-1}\alpha(1-\beta_Z)(1-\phi) + (1-\mu)\varepsilon_Z\right)\left[\mu\xi\alpha\beta_Y\overline{\phi} + \mu\varepsilon_Y\right] \end{aligned} \tag{12.7.33}$$

$$\frac{n_Z^A}{n_Z^*} = \frac{\frac{1}{(1-\mu)}\left(\frac{L^A}{L^*}\left[\xi\alpha\beta_Y\overline{\phi} + \varepsilon_Y\right] - \frac{K^A}{K^*}\left[\frac{\xi}{\xi-1}\alpha(1-\beta_Y)(1-\overline{\phi}) + \varepsilon_Y\right]\right)}{DEN2 - DEN1} \tag{12.7.34}$$

with

$$DEN1 = \left(\frac{\xi}{\xi-1}\alpha(1-\beta_Y)(1-\overline{\phi}) + \varepsilon_Y\right)[\xi\alpha\beta_Z\phi + \varepsilon_Z]$$

$$DEN2 = \left(\frac{\xi}{\xi-1}\alpha(1-\beta_Z)(1-\phi) + \varepsilon_Z\right)\left[\xi\alpha\beta_Y\overline{\phi} + \varepsilon_Y\right]$$

Solving (12.7.29) for n_Z^A/n_Z^*:

$$\frac{n_Z^A}{n_Z^*} = \frac{\frac{K^A}{K^*} - \frac{n_Y^A}{n_Y^*}\left[\mu\xi\alpha\beta_Y\overline{\phi} + \mu\varepsilon_Y\right]}{(1-\mu)\xi\alpha\beta_Z\phi + (1-\mu)\varepsilon_Z} \tag{12.7.35}$$

Plugging into L^A/L^*:

$$\begin{aligned}\frac{L^A}{L^*} = {} & \frac{n_Y^A}{n_Y^*}\left[\mu\frac{\xi}{\xi-1}\alpha(1-\beta_Y)(1-\overline{\phi}) + \mu\varepsilon_Y\right] \\ & - \frac{n_Y^A}{n_Y^*}\frac{\left[\mu\xi\alpha\beta_Y\overline{\phi} + \mu\varepsilon_Y\right]((1-\mu)\frac{\xi}{\xi-1}\alpha(1-\beta_Z)(1-\phi) + (1-\mu)\varepsilon_Z)}{(1-\mu)\xi\alpha\beta_Z\phi + (1-\mu)\varepsilon_Z} \\ & + \frac{K^A}{K^*}\frac{(1-\mu)\frac{\xi}{\xi-1}\alpha(1-\beta_Z)(1-\phi) + (1-\mu)\varepsilon_Z}{(1-\mu)\xi\alpha\beta_Z\phi + (1-\mu)\varepsilon_Z}\end{aligned} \tag{12.7.36}$$

Rearranging

$$\begin{aligned}\frac{L^A}{L^*} = {} & \frac{n_Y^A}{n_Y^*}\mu\left[\frac{\xi}{\xi-1}\alpha(1-\beta_Y)(1-\overline{\phi}) + \varepsilon_Y\right] \\ & - \frac{n_Y^A}{n_Y^*}\frac{\mu\left[\xi\alpha\beta_Y\overline{\phi} + \varepsilon_Y\right](\frac{\xi}{\xi-1}\alpha(1-\beta_Z)(1-\phi) + \varepsilon_Z)}{\xi\alpha\beta_Z\phi + \varepsilon_Z} \\ & + \frac{K^A}{K^*}\frac{\frac{\xi}{\xi-1}\alpha(1-\beta_Z)(1-\phi) + \varepsilon_Z}{\xi\alpha\beta_Z\phi + \varepsilon_Z}\end{aligned} \tag{12.7.37}$$

and multiplying with $[\xi\alpha\beta_Z\phi + \varepsilon_Z]$ and rearranging leads to

$$\begin{aligned}& \frac{L^A}{L^*}[\xi\alpha\beta_Z\phi + \varepsilon_Z] - \frac{K^A}{K^*}\left(\frac{\xi}{\xi-1}\alpha(1-\beta_Z)(1-\phi) + \varepsilon_Z\right) \\ & = \frac{n_Y^A}{n_Y^*}\mu\left[\frac{\xi}{\xi-1}\alpha(1-\beta_Y)(1-\overline{\phi}) + \varepsilon_Y\right][\xi\alpha\beta_Z\phi + \varepsilon_Z]\end{aligned}$$

$$-\frac{n_Y^A}{n_Y^*}\left(\mu\left[\xi\alpha\beta_Y\overline{\phi}+\varepsilon_Y\right]\left(\frac{\xi}{\xi-1}\alpha(1-\beta_Z)(1-\phi)+\varepsilon_Z\right)\right) \quad (12.7.38)$$

The solution for n_Y^A/n_Y^* yields:

$$\frac{n_Y^A}{n_Y^*}=\frac{\left(\frac{L^A}{L^*}\left[\xi\alpha\beta_Z\phi+\varepsilon_Z\right]-\frac{K^A}{K^*}\left[\frac{\xi}{\xi-1}\alpha(1-\beta_Z)(1-\phi)+\varepsilon_Z\right]\right)}{\mu\left[DEN1-DEN2\right]} \quad (12.7.39)$$

The results for $\frac{n_Y^A}{n_Y^*}$ and $\frac{n_Z^A}{n_Z^*}$ are symmetrical: The first summand in the denominator equals the second summand of the denominator in the other equation.

12.8 EU-Markets Susceptible to Ex Ante Regulation

Markets identified to be susceptible to ex ante regulation by European Commission in accordance with the regulatory framework for electronic communications networks and services (Source: EC (2007c) - see also Sects. 8.1.6 and 8.2.2)

1. **Retail level:** Access to the public telephone network at a fixed location for residential and non-residential customers.
2. **Wholesale level:** Call origination on the public telephone network provided at a fixed location.
3. Call termination on individual public telephone networks provided at a fixed location.
4. Wholesale (physical) network infrastructure access (including shared or fully unbundled access) at a fixed location.
5. Wholesale broadband access.
6. Wholesale terminating segments of leased lines, irrespective of the technology used to provide leased or dedicated capacity.
7. Voice call termination on individual mobile networks.

12.9 Regulatory Price Instruments

Following Dewenter and Haucap (2007), Noam (2002), Vogelsang (2003) the following pricing schemes can be used to regulate access and interconnection prices in network industries, mainly telecommunications:

- **Lump-sum**: A fixed payment.
- **Average costs pricing:** Based on historic or future-looking prices. Problems: (1) Averages are above marginal costs and may prevent potentials competitors to enter. (2) Average usage does not take into consideration peak usage, which is an important factor to determine the necessary network capacity.

- **Efficient component pricing (also ECP-rule or Baumol-Willig rule):** The entrant has to compensate the incumbent for the loss of net revenue that may be caused by entry. The optimal access price equals the marginal cost of providing the network service plus the profit foregone when interconnecting a rival instead of selling retail services. The ECP focus on productive efficiency, encourages investment in bypass facilities and innovation to overcome bottlenecks. Only ISPs which are at least as efficient as the incumbent monopolist would enter the market. But the ECP-rule also leads to allocative inefficiencies, if incumbent's retail prices are not held down to competitive levels. ECP was discussed with telecommunication regulation, but was not used by regulation authorities.
- **Incremental cost pricing:** Incremental costs include marginal costs and the part of fixed costs which can be attributed to a specific service or to a network element (e.g. the local loop). Accordingly pricing rules are also called Total Element Long Run Incremental Costs (TELRIC) or Total Services Long Run Incremental Costs (TSLRIC). Remaining costs are 'common costs'.
- **Marginal cost pricing:** Also called incremental cost pricing. In fully competitive markets, marginal cost pricing is optimal in terms of welfare and efficiency. Main problems are the difficulties to determine marginal costs in reality and the implied assumption that no further investments to build up the network are necessary. Otherwise marginal cost pricing would lead to a deficit of the natural monopoly.
- **Price caps:** Price caps give incentives for cost reduction. They can be adapted by productivity and / or inflation expectations.
- **Ramsey pricing:** Also called inverse-elasticity rule. Fixed costs are allocated among all customers according to the inverse of their demand elasticity. So customers whose demand function is price inelastic have to bear the heaviest burden. The welfare optimising principle behind this states that the behaviour of the customers is at least affected. In order that this price discrimination works arbitrage between the customer groups must be prevented. In reality Ramsey pricing is not used because the information requirements on costs and elasticities are burdensome. Furthermore Ramsey prices are criticized to peg the monopoly.
- **Two-part tariffs:** Combines a flat and variable charge, e.g. for provision and usage.
- **Wholesale pricing (also called retail-minus approach):** Determining the difference between retail and wholesale prices (margin) and to set a discount for wholesale prices. Problem is that allocative inefficiencies remain when retail prices are not regulated.

References

Adorno, T. W., & Horkheimer, M. (1969). *Dialektik der Aufklärung*. Frankfurt am Main: S. Fischer Verlag.

Aghion, P., Bloom, N., Blundell, R., Griffith, R., & Howitt, P. (2002). Competition and innovation: An inverted U relationship. *T*he Quarterly Journal of Economics, 120(2), 701–728.

Allweyer, T., Besthorn, T., & Schaaf, J. (2004). Digitale Ökonomie und struktureller Wandel. In *Economics* (Vol. 43). Deutsche Bank Research.

Amiti, M., & Wei, S.-J. (2005). Fear of service outsourcing: Is it justified. *Economic Policy, 20*(42), 307–347.

Ang, S., & Straub, D. (2002). Costs, transaction-specific investments and vendor dominance of the marketplace: The economics of is outsourcing. In R. Hirschheim, A. Heinzl, & J. Dibbern (Eds.), *Information systems outsourcing: Enduring themes, emergent patterns and future directions* (pp. 47–76). Berlin: Springer.

Antras, P. (2003). Firms, Contracts, and Trade Structure. *The Quartely Journal of Economics, 118*(4), 1375–1418.

Arino, M. (2006). Pluralism in digital broadcasting: Myths, realities and the boundaries of EU action. In B. Preissl & J. Mueller (Eds.), *Governance of communication networks* (pp. 273–294). Heidelberg: Physica.

Ark, B. V., Inklaar, R., & Mcguckin, R. (2002). *"Changing Gear" Productivity, ICT and Services Industries: Europe and the United States*. In christensen, J.F. and Maskell, P., editors, the Industrial Dynamics of the New Digital Economy, pages 56–102. Edward Elgar, Cheltenham, UK

Ark, B. V., & Piatkowski, M. (2004). *Productivity, Innovation and ICT in Old and New Europe*. Version: 2004. Retrieved December 4, 2007, from http://www.ggdc.net/pub/online/gd69(online).pdf.

Armstrong, M. (1998). Network interconnection in telecommunication. *The Economic Journal, 108*(448), 545–564.

Armstrong, M. (2002). The theory of access pricing and interconnection. In M. E. Cave, S. K. Majumdar, & I. Vogelsang (Eds.), *Handbook of telecommunications economics – volume 1 – structure, regulation and competition* (pp. 295–383). Amsterdam: Elsevier.

Armstrong, M. (2006). *Competition in Two-Sided Markets. The RAND Journal of Economics, 37*(3), 668–691

Arndt, S. W., & Kierzkowski, H. (2001). Introduction. In S. W. Arndt & H. Kierzkowski (Eds.), *Fragmentation: New production patterns in the world economy* (pp. 1–16). New York: Oxford University Press.

Arora, A., & Gambardella, A. (2004). The globalization of the software industry: Perspectives and opportunities for developed and developing countries. In *NBER Working Paper* 10538.

Arvanitis, S. (2004). Information technology, workplace organisation, human capital and firm productivity: Evidence for the Swiss economy. In *The economic impact of ICT*. Paris: OECD.

Atrostic, B. K., & Nguyen, S. (2005). *Computer investment, computer investment und productivity*. U.S. Census Bureau, CES The Center of Economic Studies (05-001).

Aubert, B. A., Patry, M., & Rivard, S. (2002). Managing IT outsourcing risk: Lessons learned. In R. Hirschheim, A. Heinzl, & J. Dibbern (Eds.), *Information systems outsourcing: Enduring themes, emergent patterns and future directions* (pp. 155–176). Berlin: Springer.

Autor, D. H., Levy, F., & Murnane, R. J. (2003). The skill content of recent technology change: An empirical exploration. *Quartely Journal of Economics, 118*(4), 1279–1333.

Baake, P., Kamecke, U., & Wey, C. (2007). Efficient regulation of dynamic telecommunications markets and the new regulatory framework in europe. In J. Haucap & R. Dewenter (Eds.), *Access pricing: Theory and practice* (pp. 373–420). Amsterdam: Elsevier.

Bacchetta, M., Matto, A., Schuknecht, L., Wagerand, H., & Wehrens, M. (1998). *Electronic commerce and the role Of the WTO*. Geneva: WTO.

Barbet, P., & Coutinet, N. (2001). *Measuring the digital economy: US and European perspective*. Issue report for STAR (Socio-Economic Trends Assessment for the Digital Revolution) program, Information Society Technologies Program of European Union.

Barfield, C. E. (2001). Electronic commerce and the GATS negotiations. In *Paper prepared for the Internet, Economic Growth and Globalization Conference Duisburg/Berlin, August 8-10 2001*.

Barnett, M. L., Starbuck, W. H., & Pant, P. N. (2003). Which dreams come true? Endogeneity, industry structure and forecasting accuracy. *Industrial and Corporate Change, 21*(4) 653–672.

Bartel, A. P., Ichniowski, C., & Shaw, K. L. (2005). How does information technology really affect productivity? Plant-level comparisons of product innovation, process improvement and worker skills. *The Quarterly Journal of Economics, 122*(4), 1721–1758.

Basu, S., & Fernald, J. (2007). Information and communication technology as a general-purpose technology: Evidence from US industry data. *German Economic Review – Special Issue on Productivity Growth, 8*(2), 146–173.

Basu, S., Fernald, J. G., Oulton, N., & Srinivasan, S. (2003). The case of the missing productivty growth: Or, does information technology explain why productivity accelerated in the US but not the UK? In *NBER Working Paper* 10010.

Baudchon, H., & Brossard, O. (2003). Definitions and Measures of ICT Impact on Growth: What is Really at Stake? In *Documents de Travail* Vol. 1. Observatoire Francais des Conjonctures Economiques (OFCE).

Belleflamme, P. (2005). Versioning in the information economy: Theory and applications. *CESinfo Economic Studies, 51*(2–3), 329–358.

Bennekoh, M., & Holznagel, B. (2005). Auswirkungen der TK-Regulierung auf die Internetmärkte dargestellt am Beispiel von Voice over IP. In *Arbeitsbericht* Vol. 39. Kompetenzzentrum Internetökonomie und Hybridität Universität Münster.

Berg, I., Eickers, G., Eilers, T., Fraenkl, C., Huber, B., Knauer, P., Kurth, M., Lennertz, H., Neumann, K.-H., Spoerr, E., & Stoeber, H. (2006). *Rahmenbedingung der Zusammenschaltung IP-basierter Netze - Abschlussbericht für Bundesnetzagentur für Elektrizität, Gas, Telekommunikation, Post und Eisenbahnen*. Version: 2006. Retrieved July 19, 2009, from http://www.bundesnetzagentur.de/media/archive/8287.pdf.

Berkey, J. O. (2002). *A Framework Agreement for Electronic Commerce Regulation unter the GATS*. Version: 2002. Retrieved April 27, 2006, from http://www.cid.harvard.edu/cidtrade/Papers/eserv_frame.pdf.

Bhagwati, J. N. (1985). Splintering and disembodiment of services and developing nations. *The world economy, 7*(2), 133–144.

Bhagwati, J. N. (2007). *In defense of globalization*. Oxford: Oxford University Press.

Bielig, A., & Haase, H. (2005). Ökonomische Implikationen der Patentierung von Computersoftware. *List Forum für Wirtschafts- und Finanzpolitik, 31*(1), 55–75.

de Bijl, P. W. J., & Peitz, M. (2006). Access Regulation and the Adoption of VoIP (Working paper presented at 17th Regional European ITS Conferences, Amsterdam, 22–24 August 2006).

Bitkom. (2007). *Grafische Online-Werbung wächst um 68 Prozent*. Version: 2007. Retrieved November 11, 2007, from http://www.bitkom.de/Default_48398.aspx.

Blankert, C. B., Knieps, G., & Zenhäusern, P. (2006). Regulation of New Markets in Telecommunications? Market dynamics and shrinking monopolistic bottlenecks (Working paper presented at 17th Regional European ITS Conferences, Amsterdam, 22–24 August 2006).

Blum, B. S., & Goldfarb, A. (2005). *Does the Internet Defy the Law of Gravity? Journal of International Economics, 70*(2), 384–405

Boes, A. (2005). Auf dem Weg in die Sackgasse? Internationalisierung im Feld Software und IT-Services. In A. Boes & M. Schwemmle (Eds.), *Offshoring und Internationalisierung im IT-Sektor* (pp. 13–65). Hamburg: VSA-Verlag.

Bomsel, O., & Geffroy, A.-G. (2006). DRM, innovation and creation. *Communications & Strategies, 62*(2), 35–47

Bonnisseau, J.-M., & Jamin, A. (2004). General equilibrium theory and increasing returns: An alternative to the survival assumption. In *Working Paper Series*. Universite Pantheon-Sorbonne.

Bosch, G. (1998). 'Jobless Growth'? Die Auswirkung der neuen Informations- und Kommunikationstechnologien auf die Beschäftigung. In *Arbeit - Zeitschrift für Arbeitsforschung, Arbeitsgestaltung und Arbeitspolitik* 4.

Botsch, M., & Stiroh, K. (2007). Information technology and productivity growth in the 2000s. *German Economic Review – Special Issue on Productivity Growth, 8*(2), 255–280.

Bourreau, M. (2005). Comment on: Martin Peitz and Patrick Waelbroeck: 'An Economist's Guide to Digital Music'. *CESinfo Economic Studies, 51*(2–3), 429–434.

Brandt, N. (2002). Market Power, Economies of Scale and the Role of Knowledge for Economic Growth. In *Dissertation des Fachbereichs Wirtschaftswissenschaften der Universität Hamburg*.

Bresnahan, T.F., & Trajtenberg, M. (1995). General purpose technologies: Engines of growth. *Journal of Econometrics, 65*, 83–108.

Briglauer, W. (2007). Market analysis under the new european communications framework – some conceptual issues. In P. J. Welfens & M. Weske (Eds.), *Digital economic dynamics* (pp. 63–90). Berlin: Springer.

Briglauer, W., & Reichinger, K. (2008). Changes of contestability in communications – a sector-specific approach. *Intereconomics, 43*(1), 51–63.

Brynjolfsson, E. (1994). *An Incomplete Contracts Theory of Information, Technology and Organization*. Version: 1994. Retrieved July 19, 2009, from http://ccs.mit.edu/papers/CCSWP126/CCSWP126.html.

Brynjolfsson, E., & Hitt, L. M. (2000). Beyond computation: Information technology, organizational transformation and business performance. *Journal of Economic Perspectives, 14*(4), 23–48.

Brynjolfsson, E., & Hitt, L. M. (2003). Computing productivity: Firm-level evidence. *The Review of Economics and Statistics, 85*(4), 793–808.

Bundesnetzagentur. (2005). *Technical regulation: Requirements for implementing statutory Telecommunications monitoring operations (TR TKÜ)*. Version: 2005. Retrieved July 19, 2009, from http://www.bundesnetzagentur.de/media/archive/5287.pdf.

BVH. (2007). *Entwicklung des E-Commerce in Deutschland (Charts zur Studie 'Distanzhandel in Deutschland 2007' von bvh/TNS-Infratest)*. Version: 2007. Retrieved July 19, 2009, from http://www.versandhandel.org/uploads/media/2007-10-29_Charts_E-Commerce_Deutschland_02.pdf.

Calvert, J. B. (2000). *The Electromagnetic Telegraph*. Version: 2000. Retrieved December 4, 2007, from http://www.du.edu/~jcalvert/tel/morse/morse.htm.

Carley, K. M. (2000). Organizational change and digital economy: A computational organization science perspective. In E. Brynjolfsson & B. Kahin (Eds.), *Understanding the digital economy* (pp. 325–351). Cambridge: MIT.

Carlton, D. W., & Picker, R. C. (2007). Antitrust and regulation. In *NBER Working Paper* 12902.

Carr, N. G. (2005). The end of corporate computing. *MIT Sloan Management Review, 46*(3), 66–73.

Castells, M. (2004). *Der Aufstieg der Netzwerkgesellschaft - Das Informationszeitalter I*. Opladen: Leske + Budrich.

Chinn, M. D., & Fairlie, R. W. (2006). ICT use in the developing world: An analysis of differences in computer and internet penetration. In *NBER Working Paper* 12382.

Chomsky, N., & Herman, E. S. (1988). *Manufacturing Consent – the political economy of the mass media*. New York: Pantheon Books.

Christou, G., & Simpson, S. (2007). *Gaining a stake in global internet governance – the EU, ICANN and Strategic norm manipulation.* (Vol. 22). 147–164.

Church, J., & Gandal, N. (2005). Platform competition in telecommunications. In M. E. Cave, S. K. Majumdar, & I. Vogelsang (Eds.), *Handbook of telecommunications economics – volume 2 - technology evolution and the internet* (pp. 117–153). Amsterdam: Elsevier.

Coase, R. H. (1937). The nature of the Firm. *Economica, 4*, 386–405.

Comino, S., Menenti, F. M., & Parisi, M. L. (2005). *From Planning to Mature: On the Determinants of Open Source Take Off.* Version: 2005. Retrieved July 19, 2009, from http://econwpa.wustl.edu/eps/io/papers/0507/0507006.pdf.

Crawford, J.-A., & Fiorentino, R. V. (2005). The changing landscape of regional trade agreements. In *Discussion paper* (Vol. 8). Geneva: World Trade Organization.

Creber, C. (2004). Die Bedeutung von Open Source in der Geschäftsstrategie von IBM. In A. Picot & S. Doeblin (Eds.), *Open Source* (pp. 111–116). Bonn: Hüthig.

Dachs, B., Meijers, H. H., & Welfens, P. J. (2007). *Internationalisation of European ICT activities – a project carried out for the European commission and institute for prospective technological studies (IPTS).* Berlin: Springer.

Darrell, D., & Sonnenschein, H. (1989). Arrow and general equilibrium theory. *Journal of Economic Literature, 27*(2), 565–598.

Devescovi, F. (2003). *Principi de Economia della Televisione.* Milano: Edizioni Angelo Guerini e Associati SpA.

Dewenter, R., & Haucap, J. (2004). *Die Liberalisierung der Telekommunikationsbranche in Deutschland. Zeitschrift für Wirtschaftspolitik, 53*(3), 374–393.

Dewenter, R., & Haucap, J. (2007). Access pricing: An introduction. In J. Haucap & R. Dewenter (Eds.), *Access pricing: Theory and practice* (pp. 1–38) Amsterdam: Elsevier.

Dibbern, J., & Heinzl, A. (2002). Outsourcing of information systems in small and medium sized enterprises: A test of a multi-theoretical causal model. In R. Hirschheim, A., Heinzl, & J. Dibbern (Eds.), *Information systems outsourcing: Enduring themes, emergent patterns and future directions* (pp. 77–99). Berlin: Springer.

Dixit, A. (1980). The role of investment in entry-deterrence. *The Economic Journal, 90*, 95–106.

DMCA. (1998). *The Digital Millennium Copyright Act of 1998.* Version: 1998. Retrieved July 19, 2009, from http://www.copyright.gov/legislation/dmca.pdf.

Dobbs, I. M. (2003). Demand, cost elasticities and pricing benchmarks in the hypothetical monopoly test: The consequences of a simple SSNIP. *Applied economics letters, 10*, 545–548.

Dunning, J. H. (1992). *Multinational enterprises and the global economy.* Wokingham, UK: Addison-Wesley.

Dutta, S., & Jain, A. (2004). *The Networked Readiness Index 2003–2004: Overview and Analysis Framework.* Version: 2004. Retrieved July 19, 2009, from http://www.weforum.org/pdf/Gcr/GITR_2003_2004/Framework_Chapter.pdf.

EC. (2002a). Commission guidelines on market analysis and the assessment of significant market power under the Community regulatory framework for electronic communications networks and services (SMP-Guideline). In *Official Journal of the European Communities* L 165/6.

EC. (2002b). Consolidated Version of the Treaty Establishing the European Community. In *Official Journal of the European Communities* C 325/35.

EC. (2002c). Consolidated version of the treaty on the European union. In *Official Journal of the European Communities* C 325/7.

EC. (2002d). Directive 2002/19/EC of the European parliament and of the council of 7 March 2002 on access to, and interconnection of, electronic communications networks and associated facilities (Access Directive). In *Official Journal of the European Communities* L 108.

EC. (2002e). Directive 2002/21/EC of the European Parliament and of the Council of 7 March 2002 on a common regulatory framework for electronic communications networks and services (Framework Directive). In *Official Journal of the European Communities* L 108/33.

EC. (2002f). Directive 2002/22/EC of the European parliament and of the council of 7 March 2002 on universal service and users' rights relating to electronic communications networks

and services (Universal Service Directive). In *Official Journal of the European Communities* L 108/51.
EC. (2004). *European Interoperability Framework for Pan-European eGovernment Services*. Version: 2004. Retrieved July 17, 2009, from http://ec.europa.eu/idabc/servlets/Doc?id=19529.
EC. (2005a). *The EU economy 2005 review: Rising international economic integration*. Brussels: Commission of the European Communities.
EC. (2005b). *i2010 — A European Information Society for Growth and Employment*. Brussels: Commission of the European Communities (COM(2005) 229 final).
EC. (2006a). *Ebusiness market watch*. Version: 2006. Retrieved July 17, 2009, from http://www.ebusiness-watch.org/key_reports/documents/EBR06.pdf.
EC. (2006b). *Proposal for a Council Regulation concerning the implementation of the Agreement on Duty-Free Treatment of Multi-Chip Integrated Circuits (MCPs) by amending Annex I to Regulation (EEC) No 2658/87 on the tariff and statistical nomenclature and on the Common Customs Tariff*. Version: 2006. Retrieved April 5, 2006, from http://europa.eu.int/eur-lex/lex/LexUriServ/site/en/com/2006/com2006_0061en01.pdf.
EC. (2007a). *Annual information society report 2007*. Brussels: Commission of the European Communities (COM(2007) 146).
EC. (2007b). *Audiovisual media services directive (AVMSD)*. Brussels: European Commission (Directive 2007/65/EC).
EC. (2007c). Commission recommendation on relevant product and service markets within the electronic communications sector susceptible to ex ante regulation in accordance with Directive 2002/21/EC of the European Parliament and of the Council on a common regulatory framework for electronic communications networks and services. In *Official Journal of the European Communities* L 344/65.
EC. (2008). Mitteilung der Kommission an das Europäische Parlament, den Rat, den Europaeischen Wirtschafts- und Sozialausschuss und den Ausschuss der Regionen über kreative Online-Inhalte im Binnenmarkt - Vorläufige Fassung. In *Commission Staff Working Document* SEK(2007)1710.
ECHR. (1950). *Convention for the Protection of Human Rights and Fundamental Freedoms*. Version: 1950. Retrieved July 17, 2009, from http://www.echr.coe.int/NR/rdonlyres/D5CC24A7-DC13-4318-B457-5C9014916D7A/0/EnglishAnglais.pdf.
Economides, N. (2004). *Telecommunications Regulation: An Introduction*. Version: 2004. Retrieved July 17, 2009, from http://www.stern.nyu.edu/eco/wkpapers/04-10Economides-revised.pdf.
Economides, N. (2005). The economics of the internet backbone. In M. E. Cave, S. K. Majumdar, & I. Vogelsang (Eds.), *Handbook of telecommunications economics – volume 2 – Technology evolution and the internet* (pp. 375–412). Amsterdam: Elsevier.
Economides, N., & Katsamakas, E. (2006). Two-sided competition of proprietary vs. open source technology platforms and the implications for the software industry. *Management Science, 52*(7), 1057–1071.
EEI. (2007). *European electronic invoicing final report*. Brussels: European Commission Informal Task Force on e-Invoicing.
Eicher, T. S., & Roehn, O. (2007). Sources of the German productivity demise: Tracing the effects of industry-level information and communication technology investment. *German Economic Review – Special Issue on Productivity Growth, 8*(2), 211–236.
EITO. (2005). *European information technology observatory*. Berlin: European Economic Interest Grouping (EEIG).
Erber, G., & Sayed-Ahmed, A. Offshore Outsourcing: A Global Shift in the Present IT Industry. In *Intereconomics* 03/04 (2005) (pp. 100–112).
ERG. (2007a). *ERG common position on VoIP*. European Regulators Group (56rev2).
ERG. (2007b). *ERG opinion on functional separation*. European Regulators Group (44).
ERG. (2007c). *ERG opinion on regulatory principles of NGA*. European Regulators Group (16rev2).
ERG. (2007d). *Final report on IP interconnection*. European Regulators Group (ERG(07)09).

Eric, B., & Curien, N. (2007). Internet economics, digital economics. In E. Brousseau (Ed.), *Internet and digital economics: Principles, methods and applications* (pp. 1–55). Cambridge: Cambridge University Press.

Erumban, A. A., & de Jong, S. B. (2005). Cross-country differences in ICT adoption: A consequence of Culture? In *Research Report* (Vol. 05C01). University of Groningen, Research Institute SOM (Systems, Organisations and Management).

Evans, D. S., Hagiu, A., & Schmalensee, R. (2005a). A survey of the economic role of software platforms in computer-based industries. *CESinfo Economic Studies, 51*(2–3), 189–224.

Evans, D. S., Nichols, A. L., & Schmalensee, R. (2005b). U.S. v. Microsoft: Did consumers win? *Journal of Competition Law and Economics, 1* (3), 497–539.

Evans, D. S., & Schmalensee, R. (2005). The industrial organization of markets with two-sided platforms. In *NBER Working Paper* 11603.

Falk, M. (2002). Diffusion der Informations- und Kommunikationstechnologien und die Qualifikationsstruktur der Arbeitskräfte. *Mitteilungen aus der Arbeitsmarkt– und Berufsforschung, 35*(3), 397–410.

Falk, M. (2003). Organizational change, new information and communication technologies and the demand for labor in services. In D. B. Audretsch, & P. J. J. Welfens (Eds.), *The new economy and economic growth in europe and the US* (pp. 161–189). Berlin: Springer.

Falk, M., & Seim, K. (2000). *The Impact of Information Technology on High-Skilled Labor in Services: Evidence from Firm-Level Panel Data.* Version: 2000. Retrieved July 17, 2009, from ftp://ftp.zew.de/pub/zew-docs/dp/dp5899.pdf.

Farrell, J., & Shapiro, C. (2004). Intellectual property, competition, and information technology. In H. R. Varian, J. Farrell, & C. Shapiro (Eds.), *The economics of information technology* (pp. 49–80). Cambridge: Cambridge University Press.

Faulhaber, G. R. (2005). Bottlenecks and bandwagons: Access policy in the new telecommunications. In M. E. Cave, S. K. Majumdar, & I. Vogelsang (Eds.), *Handbook of telecommunications economics – volume 2 – technology evolution and the internet* (pp. 487–516). Amsterdam: Elsevier.

Feenstra, R. C. (2004). *Advanced international trade.* Princeton: Princeton University Press.

Feenstra, R. C. (2007). *Comment on: Mary Amiti and Shang-Jin Wei, "Does Service Offshoring Lead to Job Losses? Evidence from the United States".* Version: 2007. Retrieved July 17, 2009, from http://www.nber.org/books/criws06/feenstra5-17-07comment.pdf.

Forge, S. (2004). *Open Source Software: Importance for Europe, Study for IPTS/JRC/EC.* Version: 2004. Retrieved July 19, 2009, from ftp://ftp.cordis.lu/pub/ist/docs/opensourcesoftware-report.pdf.

Fosfuri, A., Giarratana, M. S., & Luzzi, A. (2005). Firm assets and investments in open source software products. In *Working Paper* (Vol. 05–10). DRUID – Danish Research Unit for Industrial Dynamics.

Fowler, G. (2008). China tightens online-video regulations. *The Wall Street Journal*, p. 27.

Frank, L. and Hirvonen, T. (2006). Measuring the information society in europe: From definitions to description

Freedman, D. (2005). GATS and the audiovisual sector: An update. *Global Media and Communication, 1*, 124–128.

Freund, C. L., & Weinhold, D. (2003). The internet and evolving patterns of international trade. In C. E. Barfield, G. Heiduk, & P. J. Welfens (Eds.), *Internet, economic growth and globalization – perspectives on the new economy in Europe, Japan and the USA* (pp. 85–103). Berlin: Springer.

Frieden, R. (2003). Adjusting the horizontal and vertical in telecommunications regulation: A comparison of the traditional and new layered approach. *Federal Communications Law Journal, 55*(2), 207–250.

FSF. (2006). *Discussion Draft 1 of Version 3 of GNU General Public License.* Version: 2006. Retrieved July 19, 2009, from http://gplv3.fsf.org/draft.

Gabrielsson, P., & Gabrielsson, M. (2003). Globalizing internationals: Standardization of product and marketing strategies in the ICT field. In C. Wheeler, F. McDonald, & I. Greaves (Eds.),

Internationalization: Firm strategies and management. Basingstoke Hampshire: Palgrave Macmillan.
Ganslandt, M. (2008). *Intellectual Property Rights and Competition Policy*. Version: 2008. Retrieved July 19, 2009, from http://www.ifn.se/Wfiles/wp/wp726.pdf.
Garcia, A., & Horowitz, B. (2007). The potential for underinvestment in Internet security: Implications for regulatory policy. *Journal of Regulatory Economics, 31*, 37–55.
Geradin, D., & Sidak, J. G. (2005). European and American approaches to antitrust remedies and the institutional design of regulation in telecommunications. In S. K. Majumdar, I. Vogelsang, & M. E. Cave (Eds.), *Handbook of telecommunications economics – volume 2 – technology evolution and the internet* (pp. 517–553). Amsterdam: Elsevier Science Publishers B.V.
GIIC. (2000). *A GIIC Discussion Paper 2: 'E.-Commerce Taxation and Tariff Issues' prepared by The GIIC Electronic Commerce Committee*. Version: 2000. Retrieved July 17, 2009, from http://www.giic.org/pdf/taxation.pdf.
Giovannetti, E. (2004). Internet upstream connectivity and competition policy: Western Europe and Southern Africa. In M. Kagami, M. Tsuji, & E. Giovannetti (Eds.), *Information technology policy and the digital divide – lessons for developing countries* (pp. 35–61). Cheltenham, UK: Edward Elgar.
Giovannetti, E. (2005). Interconnection, differentiation and bottlenecks in the internet. *Information Economics and Policy, 14*(3), 385–404.
Goles, T., & Chin, W. W. (2002). Relational exchange theory and is outsourcing: Developing a scale to measure relationship factors. In R. Hirschheim, A. Heinzl, & J. Dibbern (Eds.), *Information systems outsourcing: Enduring themes, emergent patterns and future directions* (pp. 221–250). Berlin: Springer.
Gordon, R. J. (1999). Has the 'New Economy' rendered the productivity slowdown obsolete? In *NBER Working Paper* 7833.
Gordon, R. J. (2000a). Does the 'New Economy' measure up to the great inventions of the past? *Journal of Economic Perspectives, 14*(4), 49–74.
Gordon, R. J. (2000b). Interpreting the 'One Big Wave' in US: Long-term productivity growth. In *NBER Working Paper Series* 7752.
Gordon, R. J. (2003). Hi-Tech innovation and productivity growth: Does suppy create its own demand? In *NBER Working Paper* 9437.
Grossman, G. M., & Helpman, E. (1991). *Innovation and growth in the global economy*. Cambridge: MIT.
Grossman, G. M., & Helpman, E. (2002). Outsourcing in a global economy. In *NBER Working Paper* 8728.
Grossman, G. M., & Helpman, E. (2005). Outsourcing in a global economy. *Review of Economic Studies, 72*(1), 135–160.
Grossman, S. J., & Hart, O. D. (1986). The costs and benefits of ownership – a theory of vertical and lateral integration. *Journal of Political Economy, 94*(4), 691–719.
Gupta, A., Stahl, D. O., & Whinston, A. B. (2005). Pricing traffic on interconnected networks: Issues, approaches, and solutions. In M. E. Cave, S. K. Majumdar, & I. Vogelsang (Eds.), *Handbook of telecommunications economics – volume 2 – technology evolution and the Internet* (pp. 413–439). Amsterdam: Elsevier.
Haering, M. (2004). Open Source als Motor für Innovation, Geschäftsstrategie von Sun. In A. Picot & S. Doeblin (Eds.), *Open Source* (pp. 117–121). Bonn: Hüthig.
Haering, N. (2008). Der Download-Krieg der Ökonomen. In *Handelsblatt*.
Haltiwanger, J., & Jarmin, R. S. (2002). *Measuring the Digital Economy - Introduction*. Version: 2002. Retrieved July 17, 2009, from http://mitpress.mit.edu/books/BRYUH/02.haltiwanger.pdf.
Hargittai, E. (2003). The digital divide and what to do about it. In D. C. Jones (Ed.), *New economy handbook* (pp. 821–839). San Diego: Elsevier.
Harris, R. G. (2003). A communication based model of global production fragmentation. In S. W. Arndt & H. Kierzkowski (Eds.), *Fragmentation: New production patterns in the world economy* (pp. 52–87). New York: Oxford University Press – www.oup.com.

Hatfield, D. N., Mitchell, B. M., & Srinagesh, P. (2005). Emerging network technologies. In M. E. Cave, S. K. Majumdar, & I. Vogelsang (Eds.), *Handbook of telecommunications economics – volume 2 – technology evolution and the Internet* (pp. 29–80). Amsterdam: Elsevier Science Publishers B.V.

Haucap, J., Heimeshoff, U., & Uhde, A. (2007). Credible threats as an instrument of regulation for network industries. In P. J. Welfens & M. Weske (Eds.), *Digital economic dynamics* (pp. 171–202). Berlin: Springer.

Haucap, J., & Kruse, J. (2004). Verdrängungspreise auf Telekommunikationsmärkten? *Perspektiven der Wirtschaftspolitik, 5*(3), 337–361.

Hawkins, R. E. (2004). The economics of open source software for a competitve firm: Why give it away for free? *Netnomics, 6*, 103–117.

Heckmann, O., Schmitt, J., & Steinmetz, R. (2004). Optimizing interconnection policies. *Computer Networks, 46*, 19–39.

Heinemann, A. (2005). Compulsory licences and product integration in european competition law – assessment of the european commission´s microsoft decision. *International Review of Intellectual Property and Competition Law, 36*(1).

Helpman, E., & Krugman, P. R. (1985). *Market structure and foreign trade*. Cambridge: MIT.

Helpman, E., & Krugman, P. R. (1992). *Trade policy and market structure*. Cambridge: MIT.

Hempell, T., van Leeuwen, G., & van der Wiel, H. (2004). ICT, innovation and business performance in services: Evidence for Germany and the Netherlands. In *The economic impact of ICT*. Paris: OECD.

Henkel, J. (2004). Ökonomische und historische Aspekte. In A. Picot & S. Doeblin (Eds.), *Open Source* (pp. 19–28). Bonn: Hüthig.

Henten, A., & Vad, T. (2002). Services internationalization: Characteristics, potentials and barriers of Danish services firms. In M. Miozzo & I. Miles (Eds.), *Internationalization, technology and services* (pp. 184–226). Cheltenham, UK: Edward Elgar.

Hille, K. (2007,December 4). Die Wimax-Wette. *Financial Times Deutschland*.

Hillenbrand, T. (2007, March 13). Virtuelle Axt im Kopf. *Financial Times Deutschland (FTD)*.

Ho, K. W., & Hoon, H. T. (2003). *Service Links and Wage Inequality*. Version: 2003. Retrieved December 4, 2007, from http://nt2.fas.nus.edu.sg/ecs/pub/wp/wp0301.pdf. (301).

Hoekman, B. (2006). Liberalizing trade in services: A survey. In *World Bank Policy Research Working Paper* 4030.

Hoffman, D. L., & Novak, T. P. (2000). The growing digital divide: Implications for an open research agenda. In E. Brynjolfsson & B. Kahin (Eds.), *Understanding the digital economy* (pp. 245–261). Cambridge, Massachusetts: MIT.

Howell, B. (2004). *Information, Technology, and the Productivity Paradox: Incorporating Digital Goods into Productivity Measurement*. Version: 2004. Retrieved july 17, 2009, from http://www.uq.edu.au/economics/appc2004/Papers/cs6A2.pdf.

Hughes, A., & Scott-Morton, M. S. (2005). ICT and productivity growth: the paradox resolved? In *Working Paper Series* (Vol. 316). ESRC Centre for Business Research, University of Cambridge.

Huppertz, M.-T. (2004). The role of IPR for the software industry: A changing landscape? In *Patents, innovation and economic performance: OECD conference proceedings* (Vol. 13, pp. 295–298). Paris: OECD.

Huws, U., Dahlmann, S., & Flecker, J. (2004). Outsourcing of ICT and related services in the EU: A status report. Dublin, Ireland: European Foundation for the Improvement of Living and Working Conditions.

Ida, T., & Ueda, M. (2004). The interconnection and pricing of the internet. In M. Kagami, M. Tsuji, & E. Giovannetti (Eds.), *Information technology policy and the digital divide – lessons for developing countries* (pp. 282–307). Cheltenham, UK: Edward Elgar.

IDATE. (2007). *2007 digiworld yearbook*. Montpellier, France: Institut de l'Audiovisuel et des Télécommunications en Europe.

IFPI. (2007a). *Digital Music Report 2007*. Version: 2007. Retrieved July 17, 2009, from http://www.ifpi.org/content/library/digital-music-report-2007.pdf.

IFPI. (2007b). *Music Market Data H1 2007*. Version: 2007. Retrieved September 17, 2009, from http://www.ifpi.org/content/library/musicmarketsalesdata2007.pdf.

Inklaar, R., O'Mahony, M., & Timmer, M. (2005). ICT and Europe's productivity performance: Industry-level growth account comparisons with the united states. *Review of Income and Wealth, 51*(4), 505–536.

Inklaar, R., & Timmer, M. P. (2007). Of yeast and mushrooms: Patterns of industry-level productivity growth. *German Economic Review – Special Issue on Productivity Growth, 8*(2), 174–187.

Inklaar, R., Timmer, M. P., & Ark, B. V. (2006). *Mind the gap! International comparisons of productivity in services and goods production. German Economic Review-Special Issue on Productivity, 8*(2), 281–307.

IPTS. (2001). Future Bottlenecks in the Information Society: Report to the European Parliament.

ITU. (2005). *Measuring Digital Opportunity*. Version: 2005. Retrieved July 19, 2009, from http://www.itu.int/itu-wsis/2005/DOIV2.pdf.

ITU. (2006a). *digital.life - ITU Internet Report*. Version: 2006. Retrieved July 17, 2009, from http://www.itu.int/osg/spu/publications/digitalife/docs/digital-life-web.pdf.

ITU. (2006b). Interconnection in an NGN environment – background paper (advanced draft).

ITU, & UNCTAD. (2007). *World Information Society Report - Beyond WSIS*. Version: 2007. Retrieved July 17, 2009, from http://www.itu.int/osg/spu/publications/worldinformationsociety/2007/WISR07_full-free.pdf.

Jahn, E., & Prüfer, J. (2006). Interconnection and Competition among Asymmetric Networks in the Internet Backbone Market. In *Discussion Paper* (Vol. 122). Tilburg University, Center for Economic Research.

Jansson, J. O. (2006). *The economics of services*. Cheltenham, UK: Edward Elgar.

Jayatilaka, B. (2002). IT sourcing – a dynamic phenomenon: Forming an institutional theory perspective. In R. Hirschheim, A. Heinzl, & J. Dibbern (Eds.), *Information systems outsourcing: enduring themes, emergent patterns and future directions* (pp. 100–130). Berlin: Springer.

Johanson, J., & Vahlne, J.-E. (1977). The internationalization process of the firm – a model of knowledge development and increasing foreign market commitments. *Journal of International Business Studies, 8*(1), 23–32.

Jones, R. W., & Kierzkowski, H. (2001). A framework for fragmentation. In S. W. Arndt & H. Kierzkowski (Eds.), *Fragmentation: New production patterns in the world economiy* (pp. 17–34). New York: Oxford University Press.

Jorgenson, D. W. (2001). Information technology and the U.S. economy. *American Economic Review, 91*(1), 1–32.

Jorgenson, D. W. (2005). Information technology and the G7 economies. *Rivista di Politica Economica, 95*(I–II), 25–56.

Jorgenson, D. W., & Vu, K. (2007). Information technology and the world growth resurgence. *German Economic Review – Special Issue on Productivity Growth, 8*(2), 125–145.

Jullien, B. (2006). Two-sided markets and electronic intermediaries. In G. Illing & M. Peitz (Eds.), *Industrial organization and the digital economy* (pp. 272–303). Cambridge: MIT.

Kaplan, J. (2005). *Roadmap for open ICT ecosystems*. Open ePolicy Group at Berkman Center for Internet and Society at Harvard Law School.

Kapolow, L., & Shapiro, C. (2007). Antitrust. In *NBER Working Paper* 12867.

Kariyawasam, R. (2007). *International economic law and the digital divide*. Cheltenham, UK: Edward Elgar.

Kathuria, R. (2004). *Trade in Telecommunication Services: Opportunities and Constraints*. Version: 2004. Retrieved March 19, 2006, from http://www.icrier.org/wp149.pdf.

Kennet, D. M., & Ralph, E. K. (2007). Efficient interconnection charges and capacity-based pricing. *International Economics and Economic Policy (IEEP), 4*, 135–158.

Kern, T., & Willcocks, L. P. (2002). Service provision and the net: Risky application sourcing? In R. Hirschheim, A. Heinzl, & J. Dibbern (Eds.), *Information systems outsourcing: Enduring themes, emergent patterns and future directions* (pp. 511–534). Berlin: Springer.

Kiessling, J. (2006). *Institutions and ICT adoption*. Version: 2006. Retrieved July 17, 2009, from http://www.ne.su.se/paper/wp06_07.pdf.

Kirkegaard, J. F. (2004). Offshore outsourcing – much ado about what? In *CESifo Forum* 2/2004 (pp. 22–29).

Klein, H. K. (2002). On the theorectical foundations of current outsourcing research. In R. Hirschheim, A. Heinzl, & J. Dibbern (Eds.), *Information systems outsourcing: Enduring themes, emergent patterns and future directions* (pp. 24–44). Berlin: Springer.

Kling, R., & Lamb, R. (2000). IT and organizational change in digital economies: A sociotechnical approch. In E. Brynjolfsson & B. Kahin (Eds.), *Understanding the digital economy* (pp. 295–324). Cambridge: MIT.

Klodt, H. (2003). Die neue Ökonomie: Erscheinungsformen, Ursachen, und Auswirkungen. In *Kieler Studien* (Vol. 321). Berlin: Springer.

Knieps, G. (2003). Der disaggregierte Regulierungsansatz der Netzökonomie. In G. Knieps & G. Brunekreeft (Eds.), *Zwischen Regulierung und Wettbewerb - Netzsektoren in Deutschland* (pp. 9–24). Heidelberg: Physica.

Kocsis, V. (2007). Access pricing under cost asymmetry in telecommunications. In J. Haucap & R. Dewenter (Eds.), *Access pricing: Theory and practice* (pp. 149–176). Amsterdam: Elsevier.

Kocsis, V., & de Bijl, P. W. J. (2007). Network neutrality and the nature of competition between network operators. In *International Economics and Economic Policy (IEEP), 4*, 159–184.

Krajewski, M. (2005). *Auswirkungen des GATS auf Instrumente der Kulturpolitik und Kulturförderung in Deutschland - Rechtsgutachten erstellt im Auftrag der Deutschen UNESCO-Kommission*. Version: 2005. Retrieved July 17, 2009, from http://www.unesco.de/c_arbeitsgebiete/kkv_gutachten.pdf.

Krancke, J. (2003). Internationaler Handel mit Kommunikationsdienstleistungen: Anforderungen an ein multilaterales Regelwerk und die Reform des GATS. In *Kieler Studien* (Vol. 324). Berlin: Springer.

Kuropka, D., & Weske, M. (2007). Software development dynamics: Current trends and future developments. In P. J. Welfens & M. Weske (Eds.), *Digital economic dynamics* (pp. 5–24). Berlin: Springer.

Lacity, M. C., & Willcocks, L. P. (2001). *Global information technology outsourcing*. Chichester, UK: Wiley.

Laffont, J.-J., Marcus, S., Rey, P., & Tirole, J. (2003). Internet interconnection and the off-net-cost pricing principle. *RAND Journal of Economics, 34*(2), 370–390.

Lauri, F., & Hirvonen, T. (2006). *Measuring the Information Society in Europe: from Definitions to Description*. Version: 2006. Retrieved December 4, 2007, from http://www.ersa.org/ersaconfs/ersa06/papers/764.pdf.

Lawson-Borders, G. (2006). *Media organizations and convergence*. Mahwah, New Jersey: Lawrence Erlbaum Associates.

Leal-Arcas, R. (2003). United we Stand, Divided we Fall - The European Community and its Member States in the WTO Forum: towards greater Cooperation on Issues of Shared Competence? *European Political Economcy Review, 1*, 65–79.

Leamer, E. E., & Storper, M. (2001). The economic geography of the Internet age. In *NBER Working Paper* 8450.

Lee, J.-N., Huynh, M. Q., Kwok, R. C.-W., & Pi, S.-M. (2002). Current and future directions of iS outsourcing. In R. Hirschheim, A. Heinzl, & J. Dibbern (Eds.), *Information systems outsourcing: Enduring themes, emergent patterns and future directions* (pp. 195–220). Berlin: Springer.

Lee, S.-Y. T., & Guo, X. J. (2004). Information and communications technology (ICT) and spillover: A panel analysis. In *Econometric society 2004 far eastern meetings*. Department of Information Systems, National University of Singapore.

Lerner, J., & Tirole, J. (2000). The simple economics of open source. In *NBER Working Paper* 7600.

Lerner, J., & Tirole, J. (2002). The scope of open source licensing. In *NBER Working Paper* 9363.

Lerner, J., & Tirole, J. (2004). The economics of technology sharing: Open source and beyond. *The Journal of Economic Perspectives, 19*(2), 99–120.

Lessig, L. (1999). *Code and other laws of cyberspace*. New York: Basic Books.

Levine, S. S. (2004). *Towards a grounded theory of collective open source innovation*. Version: 2004. Retrieved July 17, 2009, from http://www.druid.dk/uploads/tx_picturedb/ds2004-1318.pdf.

Liebowitz, S. J., & Margolis, S. E. (2002). Network effects. In M. E. Cave, S. K. Majumdar, & I. Vogelsang (Eds.), *Handbook of telecommunications economics – volume 1* (pp. 75–96). Amsterdam: Elsevier.

Linder, J. C. (2004). Transformational outsourcing. *MIT Sloan Management Review, 45*(2), 26–31.

Lipsey, R. E. (2006). Measuring international trade in services. In *NBER Working Paper* 12271.

Liu, C. Z., & Kemerer, C. F. (2007). *Standards Competition in the Presence of Digital Conversion Technology: An Empirical Analysis of the Flash Memory Card Market*. Version: 2007. Retrieved July 17, 2009, from http://www.netinst.org/Liu_07-17.pdf.

Loeffelholz, M. (2003). Von neuen Medien zu dynamischen Systemen:Eine Bestandsaufnahme zentraler Metaphern zur Beschreibung der Emergenz öffentlicher Kommunikation. In K.-D. Altmeppen & M. Karmasin (Eds.), *Medien und Oekonomie* (pp. 53–90). Wiesbaden: Westdeutscher.

Ludwig, J. (1998). *Zur Ökonomie der Medien Zwischen Marktversagen und Querfinanzierung*. Opladen: Westdeutscher.

Mackie-Mason, J., & Varian, H. R. (1993). *Some Economics of the Internet*. Version: 1993. Retrieved July 17, 2009, from http://129.3.20.41/eps/comp/papers/9401/9401001.pdf.

Mahnke, V., & Venzin, M. (2002). How do digital information good characteristics influence pace and modalities of international market entry? In *DRUID Working Paper* (Vol. 02–13). Danish Research Unit for Industrial Dynamics (DRUID).

Mann, C. L. (2005). *This Is Bangalore Calling: Hang Up or Speed Dial?* Version: 2005. Retrieved January 30, 2006, from http://www.clevelandfed.org/Research/Com2005/0115.pdf.

Marcus, S. (2006). *Interconnection of IP-based Networks - Accounting Systems and Interconnection Regimes in the USA and UK*. Version: 2006. Retrieved July 17, 2009, from http://www.bundesnetzagentur.de/media/archive/6201.pdf.

Markusen, J. R. (2002). *Multinational firms and the theory of international trade*. Cambridge, Massachusetts: MIT.

Mattoo, A., & Wunsch, S. (2004). Pre-empting protectionism in services: The WTO and oursourcing. In *Policy research working paper series from the world bank* 3237.

Mcafee, A. (2005). Will webServices really transform Collaboration? *MIT Sloan Management Review, 46*(3).

Meckl, R., & Schramm, R. (2005). Empirical evidence for a theory of international new ventures. In *Jenaer Schriften zur Wirtschaftswissenschaft*. Friedrich-Schiller-Universität Jena.

Meunier, S., & Nicolaidis, K. (2005). The European Union as a trade power. European Union Studies Association (EUSA) - Biennial Conference - 2005 (9th), March 31-April 2 (Conference Paper).

Meyer, K., & Weinert, S. (2005). The evolution of IT outsourcing: From its origins to current and future trends. In *Arbeitspapiere des Fachbereichs Wirtschafts- und Sozialwissenschaften der Bergischen Universität Wuppertal* 202.

Mindel, J. L., & Sicker, D. C. (2006). Leveraging the EU regulatory framework to improve a layered policy model for US telecommunications markets. *Telecommunications Policy, 30*, 136–148.

Miozzo, M., & Miles, I. (2002). The relation between the internationalization of services and the process of innovation: a research agenda. In M. Miozzo & I. Miles (Eds.), *Internationalization, technology and services* (pp. 15–32). Cheltenham, UK: Edward Elgar.

Müller, G., Eymann, T., & Kreutzer, M. (2003). *Telematik- und Kommunikationssysteme in der vernetzten Wirtschaft*. Wien: Oldenbourg.

Müller, U., Utz, R., Aufderheide, D., Meyer, L., & Rodenhausen, A. (2006). Die Zukunft der Internetadressierung: ICANN, DNS und alternative Systeme - kartell- und markenrechtliche

Fragen und ihr ökonomischer Hintergrund. In *Internetökonomie und Hypbridität* (Vol. 42). Universität Münster.

Monopolkommission. (2006). *Sechzehntes Hauptgutachten der Monopolkommission 2004/2005: Mehr Wettbewerb auch im Dienstleistungssektor!* Baden-Baden: Nomos Verlagsgesellschaft.

Motta, M., & Polo, M. (2005). *Competition policy: Theory and practice*. New York: Cambridge University Press.

Neary, J. P. (2004). *Monopolistic Competition and international Trade Theory*. In Brakman, S. and Heiÿdra, B.J., editors, The Monopolistic Competition Revolution in Retrospect, pages 159–184. Cambridge University Press, Cambridge.

Nefiodow, L. A. (2001). *Der Sechste Kondratief - Wege zur Produktivität und Vollbeschäftigung im Zeitalter der Information*. Sankt Augustin: Rhein-Sieg.

Netcraft.Com. (2007). *Web Server Survey August 2007*. Version: 2007. Retrieved October 3, 2007, from http://news.netcraft.com/archives/2007/08/index.html.

Noam, E. M. (2002). Interconnection practices. In M. E. Cave, S. K. Majumdar, & I. Vogelsang (Eds.), *Handbook of telecommunications economics – volume 1 – structure, regulation and competition* (pp. 387–421). Amsterdam: Elsevier.

Noam, E. (2006). Why TV regulation will become telecom regulation. In E. Richards, R. Foster, & T. Kiedrowski (Eds.), *Communications – the next decade – a collection of essays prepared for the UK office of communications* (pp. 67–72). London: Ofcom.

Nordhaus, W. D. (2007). Two centuries of productivity growth in computing. *The Journal of Economic History, 67*(1), 128–159.

Nuechterlein, J. E., & Weiser, P. J. (2005). *Digital crossroads – American telecommunications policy in the internet age*. Cambridge: MIT.

Oberholzer-Gee, F., & Strumpf, K. (2007). The effect of file sharing on record sales : an empirical analysis. In *Journal of Political Economy*, 1–42.

OECD. (2001). *Measuring productivity – OECD manual: Measurement of aggregate and industry-level productivity growth*. Paris: OECD.

OECD. (2004a). *The Implications of Convergence for Regulation of Electronic Communications*. Version: 2004. Retrieved December 4, 2007, from http://195.218.115.39/pport/pdf/337570388.pdf.

OECD. (2004b). *Information technology outlook 2004*. Paris: OECD.

OECD. (2004c). *Understanding economic growth*. Paris: OECD.

OECD. (2005). *Guide to Measuring the Information Society*. Version: 2005. Retrieved July 20, 2009, from http://www.oecd.org/dataoecd/41/12/36177203.pdf.

OECD. (2006a). *Digital Broadband Content: Digital Content Strategies and Policies*. Version: 2006. Retrieved July 20, 2009, from http://www.oecd.org/dataoecd/54/36/36854975.pdf.

OECD. (2006b). *The Impact of Substitute Services on Regulation*. Version: 2006. Retrieved July 20, 2009, from http://www.oecd.org/dataoecd/13/11/36997290.pdf.

OECD. (2006c). *The Implications of WiMax for Competition and Regulation*. Version: 2006. Retrieved July 20, 2009, from http://www.oecd.org/dataoecd/32/7/36218739.pdf.

OECD. (2006d). *Internet Traffic Exchange: Market Developments and Measurement of Growth*. Version: 2006. Retrieved July 20, 2009, from http://www.oecd.org/dataoecd/25/54/36462170.pdf.

OECD. (2006e). *Policy Considerations of VoIP*. Version: 2006. Retrieved July 20, 2009, from http://www.oecd.org/dataoecd/59/55/36316212.pdf.

OECD. (2006f). *The Policy Implications of Voice over Internet Protocol*. Version: 2006. Retrieved July 20, 2009, from http://www.oecd.org/dataoecd/14/29/36133304.pdf.

OECD. (2006g). *Rethinking Universal Service for a Next Generation Network Environment*. Version: 2006. Retrieved July 20, 2009, from http://www.oecd.org/dataoecd/59/48/36503873.pdf.

OECD. (2006h). *VoIP: Developments in the Market*. Version: 2006. Retrieved July 20, 2009, from http://www.oecd.org/dataoecd/56/24/35955832.pdf.

OECD. (2007a). *Classifying Information and Communication Technology (ICT) Services*. Version: 2007. Retrieved July 20, 2009, from http://www.oecd.org/dataoecd/39/25/38226951.pdf.

OECD. (2007b). *Fixed-mobile convergence: Market Developments and Policy Issues*. Version: 2007. Retrieved July 20, 2009, from http://www.oecd.org/dataoecd/20/26/38309911.pdf.

OECD. (2007c). *Internet Traffic Prioritisation: An overview*. Version: 2007. Retrieved July 20, 2009, from http://www.oecd.org/dataoecd/43/63/38405781.pdf.

OECD. (2007d). *Mobile Multiple Play: New Service Pricing and Policy Implications*. Version: 2007. Retrieved July 20, 2009, from http://www.oecd.org/dataoecd/18/19/38057443.pdf.

OECD. (2007e). *National broadband action plans*. Version: 2007. Retrieved July 20, 2009, from https://www.oecd.org/countrylist/0,3349,en_17642234_17642806_38711225_1_1_1_1,00.html.

OECD. (2007f). *OECD communications outlook 2007*. Paris: OECD.

OECD. (2007g). *Participative web and user-created content: Web 2.0, wikis and social networking*. Paris: OECD.

OECD. (2007h). *Policy Considerations for Audio-Visual Content Distribution in a Multiplatform Environment*. Version: 2007. Retrieved July 20, 2009, from http://www.oecd.org/dataoecd/21/41/37868139.pdf.

Ofcom. (2007a). *The Communications Market Special Report Niche ISPs*. Version: 2007. Retrieved July 20, 2009, from http://www.ofcom.org.uk/research/cm/nicheisp/niche.pdf.

Ofcom. (2007b). *The International Communication Market 2007*. Version: 2007. Retrieved July 20, 2009, from http://www.ofcom.org.uk/research/cm/icmr07/icmr07.pdf.

Oliner, S. D., & Sichel, D. E. (2000). The resurgence of growth in the late 1990s: Is information technology the story. *Journal of Economic Perspectives, 14*(4), 3–22.

Oliner, S. D., & Sichel, D. E. (2002). *Information Technology and Productivity: Where Are We Now and Where Are We Going?* Version: 2002. Retrieved September 17, 2009, from http://www.federalreserve.gov/pubs/feds/2002/200229/200229pap.pdf.

Park, Y., & Scotchmer, S. (2005). Digital rights management and the pricing of digital products. In *NBER Working Paper* 11532.

Peitz, M. (2006). Markplätze und indirekte Netzwerkeffekte. *Perspektiven der Wirtschaftspolitik, 7*(3), 317–334.

Peitz, M., & Waelbroeck, P. (2005). An economist's guide to digital music. *CESinfo Economic Studies, 51*(2–3), 359–428.

Pepall, L., Richards, D. J., & Norman, G. (2005). *Industrial organization – contemporary theory and practice*. Mason, Ohio: Thomson South-Western.

Piazolo, D. (2001). The new economy and the international regulatory framework. In *Kiel Working Paper* (Vol. 1030). Kiel Institute of World Economics.

Pilat, D. (2004). The ICT productivity paradox: Insights from micro data. In *OECD Economic Studies* 2004-1 (pp. 37–65).

Pilat, D., & Devlin, A. (2004). The diffusion of ICT in OECD economies. In *The economic impact of ICT*. Paris: OECD.

Pilat, D., & Wölfl, A. (2004). ICT production and ICT use: What role in aggregate productivity growth? In *The economic impact of ICT*. Paris: OECD.

Pilch, H. (2004). Why are software patents so trivial? In *Patents, innovation and economic performance: OECD conference proceedings* (Vol. 13, pp. 289–294). Paris: OECD.

Ponder, J. (2006). What regulatory paradigm for IP-enabled NGNs? (Presentation during ITS conference in Amsterdam August).

Ponder, J. (2007). Information society 2015: Towards increased participation. In P. J. Welfens & M. Weske (Eds.), *Digital economic dynamics* (pp. 203–210). Berlin: Springer.

Portman, R. (2005). *Opening Remarks of Ambassador Rob Portman United States Trade Representative*. Version: 2005. Retrieved April 5, 2006, from http://www.ustr.gov/assets/Document_Library/USTR_Speeches/2005/asset_upload_file465_8334.pdf.

Quah, D. (2002). *Digital Goods and The New Economy*. Version: 2002. Retrieved July 17, 2009, from http://econ.lse.ac.uk/staff/dquah/p/dp-0212hbne.pdf.

Quinn, B. (2000). Outsourcing innovation: The new engine of growth. *MIT Sloan Management Review, 41*(4), 13–28.

Rauch, J. E., & Trindade, V. (2003). Information and globalization: Wage co-movements, labor demand elasticity, and conventional trade liberalization. *American Economic Review, 93*, 775–791.

Rayna, T., & Striukova, L. (2007). Digital Rights Management: White Knight or Trojan Horse? In *Bristol Economics Discussion Papers* Vol. 07/596. University of Bristol.

Regibeau, P. (2005). A comment on Evans, Hagiu and Schmalensee. *CESinfo Economic Studies, 51*(2–3), 225–232.

Röger, W. (2002). Structural changes and the new economy in the EU and the US. In D. B. Audretsch & P. J. Welfens (Eds.), *The new economy and economic growth in Europe and the US* (pp. 7–27). Berlin: Springer.

Rochet, J.-C., & Tirole, J. (2005). *Two Sided Markets: A Progress Report. The Rand Journal of Economics, 37*(3), 645–667

Rohlfs, J. H. (2005). Bandwagon effects in telecommunications. In M. E. Cave, S. K. Majumdar, & I. Vogelsang (Eds.), *Handbook of telecommunications economics – volume 2 – technology evolution and the Internet* (pp. 79–115). Amsterdam: Elsevier.

Romer, P. (1989). Endogenous technological change. *Journal of Political Economy, 98*(5), 71–102

Roson, R. (2005). Two-sided markets: A tentative survey. *Review of Network Economics, 4*(2), 142–160

Roy, M., Marchetti, J., & Lim, H. (2006). *Services Liberalization in the New Generation of Preferential Trade Agreements (PTAs): How Much Further than the GATS?* Version: 2006. Retrieved July 20, 2009, from http://www.wto.org/english/res_e/reser_e/ersd200607_e.pdf.

Rubalcaba-Bermejo, L. (2004). The globalisation of business services and the competitiveness of European ICT services, XIV Conference RESER, 23–24 September 2004 – Castres.

Rubalcaba-Bermejo, L., & Gago-Saldana, D. (2002). National versus international effects in regional concentration of European innovative business services. In M. Miozzo & I. Miles (Eds.), *Internationalization, technology and services* (pp. 137–157). Cheltenham, UK: Edward Elgar.

Sadowski, B., & Straathof, B. (2005). *VoIP under the EU Regulatory Framework: Preventing Foreclosure?* Version: 2005. Retrieved January 16, 2008, from http://web.si.umich.edu/tprc/papers/2005/429/SadowskiStraathof2005Final.pdf.

Sander, H. (2002). Integrating eletronic commerce into the world trading system: Issues and challenges. In D. B. Audretsch & P. J. Welfens (Eds.), *The new economy and economic growth in Europe and the US* (pp. 253–270). Berlin: Springer.

Schettkat, R., & Russo, G. (2001). Structural dynamics and employment in highly industrialized economies. In P. Petit & L. Soete (Eds.), *Technology and the future of European employment* (pp. 111–143). Edward Elgar.

Schleife, K. (2006). Regional versus Individual Aspects of the Digital Divide in Germany. In *Arbeitspapiere des Instituts für Volkswirtschaftslehre Technische Universität Darmstadt* 177.

Schumpeter, J. A. (1942). *Capitalism, Socialism and Democracy*. London: Harper Perennial (republished 1984) HarperCollins Publishers.

Schwartz, P. M., & Janger, E. J. (2007). Notification of data security breaches. *Michigan Law Review, 105*(5), 913–985.

Sciadas, G. (2004). *International Benchmarking for the Information Society - A background paper for the ITU-KADO Digital Bridges Symposium*. Version: 2004. Retrieved July 17, 2009, from http://www.itu.int/osg/spu/ni/digitalbridges/docs/background/BDB-intl-indices.pdf.

SEC. (2007). *Google Inc - Registration Statement as filed with the Securities and Exchange Commission (SEC) on February 7, 2007*. Version: 2007. Retrieved July 17, 2009, from http://www.sec.gov/Archives/edgar/data/1288776/000119312507022578/ds3asr.htm.

Shapiro, C., & Varian, H. R. (1998). *Information rules*. Boston: Harvard Business School Press.

Shaw, K. (2002). By what means does information technology affect employment and wages? In N. Greenan, Y. L'Horty, & J. Mairesse (Eds.), *Productivity, inequality and the digital economy* (pp. 229–268). Cambridge: MIT.

Shy, O. (2002). *The economics of network industries*. Cambridge, UK: Cambridge University Press.

Smith, M. D., Bailey, J., & Brynjolfsson, E. (2000). Understanding digital markets: Review and assessment. In E. Brynjolfsson & B. Kahin (Eds.), *Understanding the digital economy* (pp. 99–136). Cambridge: MIT.

Spata, J. B. (2006). Rewriting U.S: Telecommunications law with an eye on Europe. In B. Preissl & J. Mueller (Eds.), *Governance of communication networks* (pp. 11–36). Heidelberg: Physica-Verlag.

Spencer, B. J. (2005). International outsourcing and incomplete contracts. In *NBER Working Paper* 11418.

Spiezia, V., & Vivarelli, M. (2002). Innovation and employment: A critical survey. In N. Greenan, Y. L'Horty, & J. Mairesse (Eds.), *Productivity, inequality and the digital economy* (pp. 101–132). Cambridge: MIT.

Standage, T. (2007). *The Victorian internet: The remarkable story of the telegraph and the nineteenth century's on-line pioneers*. Walker and Company.

Stenborg, M. (2004). Explaining Open Source. In *Discussion papers* (Vol. 947). ETLA - The Research Institute of the Finnish Economy.

Stieler, A. (1891). *Telegraphen Verbindungen - Weltkarte Abbildung Nr. 5*. Version: 1891. Retrieved July 20, 2009, from http://en.wikipedia.org/wiki/Image:1891_Telegraph_Lines.jpg.

Stiroh, K. J. (2002). Information technology and the U.S. productivity revival: What do the industry data say? *American Economic Review, 92*(5), 1559–1576.

Streel, A. D. (2007). Antitrust and sector-specific regulation in the European union: The case of electronic communications. In J. Haucap & R. Dewenter (Eds.), *Access pricing: Theory and practice* (pp. 327–368). Amsterdam: Elsevier.

Stumpf, U. (2006). Markets susceptible to ex ante regulation : Methodology and commission recommendation. *Communications & Strategies, 41*(4), 41–62.

SVR. (2007). *Das Erreichte nicht verspielen - Jahresgutachten 2007*. Reutlingen: Sachverständigenrat zur Begutachtung der gesamtwirtschaftlichen Entwicklung, Fachverlage GmbH.

Tambini, D. (2006). What citizens need to know. Digital exclusion, information inequality and rights. In E. Richards, R. Foster, & T. Kiedrowski (eds.), *Communications – the next decade – a collection of essays prepared for the UK office of communications* (pp. 112–124). London: Ofcom.

Tanaka, G. Y. (2004). *Digital deflation: The productivity revolution and how it will ignite the economy*. New York: McGraw-Hill.

Tardiff, T. J. (2006). The economics of access and interconnection charges in telecommunications. In M. A. Crew (Ed.), *International handbook on economic regulation* (pp. 278–301). Cheltenham, UK: Edward Elgar.

Thompson, R. (2002). Electronic Commerce and the WTO. In *Evian Group Compendium September 2002*.

Triplett, J. (2004). *Handbook on Hedonic Indexes and Quality Adjustments in Price Indexes. OECD - Directorate for Science, Technology and Industry*. Version: 2004. Retrieved July 20, 2009, from http://www.oecd.org/dataoecd/37/31/33789552.pdf.

Tuomi, I. (2004). Knowledge society and the new productivity paradigm: A critical review of productivity theory and the impacts of ICT. In *IPTS Working Paper*.

UN. (2002). Manual on statistics of international trade in services. In *Statistical Papers* (Vol. M 86). United Nations, European Commission, International Monetary Fund, Organisation for Economic Co-operation and Development, United Nations Conference on Trade and Development, World Trade Organization.

UN. (2008). *United nations e-Government survey 2008: From e-Government to Connected Governance*. New York: United Nations.

UNCTAD. (2005) *Core ICT Indicators*. Version: 2005. http://new.unctad.org/upload/docs/CoreICTIndicators_Eng.pdf. – online resource, lastchecked: 17.20.2009

UNCTAD. (2006). *Information Economy Report 2006 - The Development Perspective (Ref. UNCTAD/SDTE/ECB/2006/1)*. Version: 2006. Retrieved September 17, 2009, from http://www.unctad.org/en/docs/sdteecb20061_en.pdf.

USCensusbureau. (2007a). *E-Stats*. U.S. Census Bureau http://www.census.gov/eos/www/2005/2005reportfinal.pdf
USCensusbureau. (2007b). *Estimated Quarterly U.S. Retail Sales: Total and E-commerce*. Version: 2007. Retrieved October 31, 2007, from http://www.census.gov.
USTR. (2003). *The 2003 Annual Report and 2004 Trade Policy Agenda: World Trade Organization*. Washington, D.C.: Office of the United States Trade Reprensentative.
USTR. (2005). *The Dominican Republic - Central America - United States Free Trade Agreement*. Version: 2005. Retrieved April, 14, 2006, from http://www.ustr.gov/Trade_Agreements/Bilateral/CAFTA/CAFTA-DR_Final_Texts/Section_Index.html.
USTR. (2006a). *2006 Trade Policy Agenda and 2005 Annual Report of the President of the United States on the Trade Agreements Program*. Version: 2006. Retrieved April 14, 2006, from http://www.ustr.gov/assets/Document_Library/Reports_Publications/2006/2006_Trade_Policy_Agenda/asset_upload_file765_9077.pdf.
USTR. (2006b). *E-Commerce FTA Chapters*. Version: 2006. Retrieved April 14, 2006, from http://www.ustr.gov/Trade_Sectors/Telecom-E-commerce/E-Commerce_FTA_Chapters/Section_Index.html.
USTR. (2006c). *Trade Facts: NAFTA: A Strong Record of Success*. Version: 2006. Retrieved April 14, 2006, from http://www.ustr.gov/assets/Document_Library/Fact_Sheets/2006/asset_upload_file242_9156.pdf.
Vanberg, M. (2005). Network externalities and interconnection incentives. In *ZEW Discussion Papers* (Vol. 05-80). Mannheim: Zentrum für Europäische Wirtschaftsforschung.
Vanberg, M. (2007). Implications of network convergence on local access regulation in the U.S. and the EU. In *ZEW Discussion Paper* (Vol. 07-039). Mannheim: Zentrum für Europäische Wirtschaftsforschung.
Varian, H. R. (2000). Market structure in the network age. In E. Brynjolfsson & B. Kahin (Eds.), *Understanding the digital economy* (pp. 137–150). Cambridge: MIT.
Varian, H. R. (2004). Competiton and market power. In H. R. Varian, J. Farrell, & C. Shapiro, (Eds.), *The economics of information technology*. Cambridge: Cambridge University Press.
Venturini, F. (2006). ICT and Productivity Resurgence: a growth model for the Information Age (First Version), Universita' Politecnica delle Marche (I), Dipartimento di Economia.
Vernon, R. (1966). International investment and international trade in the product cycle. *Quartely Journal of Economics, 80*, 190–207.
Vickery, G., & Wunsch-Vincent, S. (2005). ICT use in services and the growth of digital content services. In *Enhancing the performance of the service sector* (pp. 205–238). Paris: OECD/RIETI.
Vogelsang, I. (2003). Price regulation of access to telecommunications network. In *Journal of Economic Literature, XLI*, 830–862.
Vogelsang, I. (2006). *Abrechnungssysteme und Zusammenschaltungsregime aus ökonomischer Sicht - Gutachten im Auftrag der Bundesnetzagentur für Elektrizität, Gas, Telekommunikation, Post und Eisenbahnen*. Version: 2006. Retrieved July 20, 2009, from http://www.bundesnetzagentur.de/media/archive/6202.pdf.
Wagner, C. (2004). Konvergenz und Trennung der Regulierung von Netz und Inhalt - Fragen an das EG-rechtliche Konzept für die Fernsehverbreitung per Satellit. In A. Scheuer & J. Gerth (eds.), *Digitale Satellitenplattformen in den USA und Europa und ihre Regulierung 9. Medienrechtskolloquium der Landesanstalt für Medien (LfM) Nordrhein-Westfalen und des Instituts für Europäisches Medienrecht e.V. (EMR), Saarbrücken/Brüssel* (Vol. 29, pp. 41–55). Landesanstalt für Medien NRW.
Walter, B. V., & Quiring, O. (2006). The transformation of media – economic and social implications. In B. Preissl & J. Mueller (Eds.), *Governance of communication networks* (pp. 244–272). Heidelberg: Physica.
Ward, M. R. (2007). *Digital Piracy*. Version: 2007. Retrieved February 11, 2008, http://papers.ssrn.com/sol3/papers.cfm?abstract_id=983200.
Waverman, L. (2006). The challenges of a digital world and the need for a new regulatory paradigm. In E. Richards, R. Foster, & T. Kiedrowski (Eds.), *Communications – the next decade – a collection of essays prepared for the UK office of communications* (pp. 158–175). London: Ofcom.

World Economic Forum, & INSEAD. (2007). *The global information technology report 2006–2007*. Basingstoke, Hampshire: World Economic Forum and INSEAD, Palgrave Macmillan.
Wegner, R., & Bachmeier, C. (2000). *Streaming Media im Business Bereich - Echtzeitverfahren für das WWW*. München: Addison-Wesley.
Welfens, P. J. J. (2001). *Interneteconomics.net*. Berlin: Springer.
Welfens, P. J. J. (2004a). Auf dem Weg in eine Europäische Informations- und Wissensgesellschaft: Probleme, Weichenstellungen, Politikoptionen. In *Discussion Paper* (Vol. 117). Wuppertal, Potsdam: European Institute for International Economic Relations (EIIW).
Welfens, P. J. J. (2004b). *Information and communication technology, regulations and growth: Theory and key policy issues (Presentation December 2004)*. Wuppertal, Potsdam: European Institute for International Economic Relations (EIIW).
Welfens, P. J. J. (2006). *Die Zukunft des Telekommunikationsmarktes - Volks-wirtschaftliche Aspekte digitaler Wirtschaftsdynamik*. Berlin: Friedrich Ebert Stiftung.
Welfens, P. J. J. (2007a). Digital social market economy – towards a new economic system. In P. J. Welfens & M. Weske (Eds.), *Digital economic dynamics* (pp. 25–62). Berlin: Springer.
Welfens, P. J. J. (2007b). Schattenregulierung als Phänomen der Telekommunikationswirtschaft. *List Forum für Wirtschafts- und Finanzpolitik, 33*(3), 179–205.
Welfens, P. J. (2007c). *Digital integration, growth and rational regulation*. Berlin: Springer.
Welfens, P. J. (2007d). *Grundlagen der Wirtschaftspolitik*. Berlin: Springer.
Welfens, P. J., & Jungmittag, A. (2002a). Europäische Telekomliberalisierung und Außenhandel: Theorie, Gravitationsansatz und Implikationen. *Jahrbücher für Nationalökonomie und Statistik, 222*, 98–111.
Welfens, P. J., & Jungmittag, A. (2002b). *Internet, Telekomliberalisierung und Wirtschaftswachstum*. Berlin: Springer.
Welfens, P. J., Jungmittag, A., & Vogelsang, M. (2007). *Innovation, Regulierung und Wirtschaftswachstum in Digital Marktwirtschaften: National und regionale Perspektiven*. Lohmar: Josef Eul.
Welfens, P. J., & Vogelsang, M. (2008). Theory and concepts. In B. Dachs, H. Meijers, & P. J. Welfens (Eds.), *Internationalisation of European ICT activities* (pp. 6–83). Berlin: Springer.
Wellenius, B., & Townsend, D. N. (2005). Telecommunications and economic development. In M. E. Cave, S. K. Majumdar, & I. Vogelsang (Eds.), *Handbook of telecommunications economics – volume 2 – technology evolution and the internet* (pp. 555–619). Amsterdam: Elsevier.
Welsum, D. V., & Vickery, G. (2005). *Potential offshoring of ICT-Intensive using occupations*. Paris: OECD.
Werbach, K. (2002). A layered model for internet policy. *Journal of Telecommunications and High-Tech Law, 1*, 37–68.
Whitehouse. (2003). *National Strategy to Secure Cyberspace*. Version: 2003. Retrieved April 23, 2008, from http://www.whitehouse.gov/pcipb/cyberspace_strategy.pdf.
Whitt, R. S. (2004). A horizontal leap forward: Formulating a new communications public policy framework based on the network layers model. *Federal Communications Law Journal, 56*(3), 587–672.
Wiese, H. (2002). Strategic trade policy in the presence of network effects. *German Economic Review, 3*, 155–183.
Williamson, O. E. (1985). *The economic institutions of capitalism*. New York: Free Press.
Willig, R. D. (2004). Competition policy for network and internet markets. In L. Paganetto (Ed.), *Knowledge economy, information technologies and growth* (pp. 137–158). Ashgate: Aldershot.
WIPO. (2005). *Annual Report 2004*. World Intellectual Property Organization.
WTO. (1986). *GATT – The General Agreement on Tariffs and Trade*. Geneva: World Trade Organization.
WTO. (1991). *Services sectoral classification list*. Geneva: World Trade Organization (MTN.GNS/W/120).
WTO. (1994). *General Agreement on Trade in Services (GATS)*. Version: 1994. Retrieved July 20, 2009, from http://www.wto.org/english/docs_e/legal_e/26-gats.pdf.

WTO. (1999). *Work programme on eletronic commerce – progress report to the general council*. Geneva: World Trade Organization (S/L/74).

WTO. (2002). *Communication from the European communities and their member states: Coverage of CPC 84 – computer and related services*. Geneva: World Trade Organisation (TN/S/W/6).

WTO. (2005a). *Council for trade in services – communication from the European communities and its member states – conditional revised offer*. Geneva: World Trade Organization (TN/S/O/EEC/Rev.1).

WTO. (2005b). *Doha Work Programme - Ministerial Declaration - Adopted on 18 December 2005*. Version: 2005. Retrieved July 20, 2009, from http://www.wto.org/english/thewto_e/minist_e/min05_e/final_text_e.htm.

WTO. (2005c). *Sixth dedicated discussion on eletronic Eommerce under the auspices of the general council on 7 and 21 November 2005 – summary by the secretariat of the issues raised*. Geneva: World Trade Organization (WT/GC/W/556).

WTO. (2005d). *Understanding the WTO*. Geneva Version: 2005. Retrieved July 20, 2009, from http://www.wto.org/english/thewto_e/whatis_e/tif_e/understanding_text_e.pdf.

WTO. (2005e). *United States – measures affecting the cross-border supply of gambling and betting services*. (Vol. ARB-2005-2/19). Geneva: World Trade Organization.

WTO. (2005f). *World trade report 2005*. Lausanne:WTO.

WTO. (2007a). *Understanding on the scope of coverage of CPC 84 – computer and related services*. (Vol. TN/S/W/60). Geneva: World Trade Organization.

WTO. (2007b). *World Trade Report 2007*. Version: 2007. Retrieved July 20, 2009, from http://www.wto.org/english/res_e/booksp_e/anrep_e/world_trade_report07_e.pdf.

Wu, I. (2004). Canada, South Korea, Netherlands and Sweden: Regulatory implications of the convergence of telecommunications, broadcasting and internet services. *Telecommunications Policy, 28*, 79–96.

Wunsch-Vincent, S. (2003). The digital trade agenda of the U.S.: Parallel tracks of bilateral, regional and multilateral liberalization. *Aussenwirtschaft, 58*(1), 7–46.

Wunsch-Vincent, S. (2005a). *Market access for digitally-delivered content products and the resulting challenges to the WTO: A US versus an EC perspective (Phd Thesis)*. Bamberg: Difo-Druck GmbH.

Wunsch-Vincent, S. (2005b). WTO, E-Commerce and information technologies: From the Uruguay round through the Doha development agenda – a report to the UN ICT task force. In *ICT Task Force Series* (Vol. 7). New York: United Nations ICT Task Force.

Yomogida, M. (2004). *Communication Costs, Producer Services, and International Trade*. Version: 2004. Retrieved July 20, 2009, from http://www.etsg.org/ETSG2004/Papers/Yomogida.pdf.

Zampetti, A. B. (2003). *WTO Rules in the Audio-Visual Sector*. Version: 2003. Retrieved August 17, 2008, from http://www.hwwa.de/Forschung/Publikationen/Report/2003/Report229.pdf.

Zhu, Z., Hsu, K., & Lillie, J. (2001). Outsourcing - a strategic move: The process and the ingredients for success. *Management Decision, 39*(5/6), 373–378.

Zimny, Z., & Mallampally, P. (2002). Internationalization of services: are the modes changing? In M. Miozzo & I. Miles (Eds.), *Internationalization, technology and services* (pp. 87–114). Cheltenham, UK: Edward Elgar.

Zoellick, R. B. (2004). *Statement of Robert B. Zoellick before the Committee on Ways and Means of the United States House of Representatives March 11, 2004*. Version: 2004. Retrieved March 14, 2007, from http://www.ustr.gov/speech-test/zoellick/2004-03-11-housewaysmeans.pdf.

GPSR Compliance

The European Union's (EU) General Product Safety Regulation (GPSR) is a set of rules that requires consumer products to be safe and our obligations to ensure this.

If you have any concerns about our products, you can contact us on ProductSafety@springernature.com

In case Publisher is established outside the EU, the EU authorized representative is:

Springer Nature Customer Service Center GmbH
Europaplatz 3
69115 Heidelberg, Germany

Batch number: 10725802

Printed by Printforce, the Netherlands